·ROYAL HORTICULTURAL SOCIETY·
GREAT BRITISH
VILLAGE SHOW

·ROYAL HORTICULTURAL SOCIETY·

GREAT BRITISH
VILLAGE SHOW

Thane Prince & Matthew Biggs

Foreword by

ALAN TITCHMARSH

Contents

8 Foreword by **Alan Titchmarsh**

10 Introductions by **Matthew Biggs** and **Thane Prince**

GROW TO SHOW FRUIT

14 Tips & Tricks:
Fruit Classes

16 *Staging the Exhibits*

18 Strawberries

19 Raspberries

20 Blackberries & Hybrids

24 Currants

26 Gooseberries

28 Cherries

29 Apricots

30 Plums

31 Damsons & Bullaces

34 Figs

35 Citrus

36 Apples

38 *A World of Apples*

40 *The RHS Judge*

42 Pears

43 *A World of Pears*

44 Quince

45 Medlars

48 Grapes

50 Cape Gooseberries

51 Rhubarb

GROW TO SHOW VEGETABLES

54 Tips & Tricks:
Vegetable Classes

56 Tomatoes

58 Sweet Peppers

59 Chilli Peppers

60 *The Urban Gardener*

62 Aubergines

63 Sweetcorn

64 Broad Beans

65 French Beans

66 Runner Beans

67 Peas

70 Cucumbers

72 Courgettes

73 Marrows

74 Squash

75 Pumpkins

76 *A Weird World of
Pumpkins & Squash*

78 *The Giant Pumpkin Grower*

80 Asparagus

81 Globe Artichokes

82 Leeks

84 Onions

86 Shallots

87 Garlic

88 Spring Onions

89 Celery

90 Celeriac

91 Florence Fennel

92 *Judging the Exhibits*

94 Carrots

96 Beetroot

97 Kohlrabi

98 Turnips

99 Parsnips

102 Jerusalem Artichokes

103 Potatoes

104 Radishes

106 Lettuces

110 Cabbages

112 *The Horticultural
Society Chairman*

114 Kale

115 Brussels Sprouts

116 Sprouting Broccoli

117 Calabrese

118 Cauliflowers

119 Chard

120 Herbs

GROW TO SHOW FLOWERS

HOMECRAFT BAKE & PRESERVE

124 Tips & Tricks:
    Flower Classes

126 Floral Displays

127 House Plants

128 *The Floral Artist
    & Show Organiser*

130 Cacti & Succulents

132 Bonsai

134 Orchids

135 Alpine Plants

136 *The Gardening Dynasty*

138 Primulas

139 Tulips

140 Irises

141 Pansies

142 Carnations & Pinks

144 Delphiniums

145 Sweet Peas

148 Roses

150 Begonias

151 Gladioli

152 Fuchsias

153 Pelargoniums

156 Dahlias

158 *The Dahlia Exhibitor*

160 Chrysanthemums

161 *Awarding the Prizes*

166 Tips & Tricks:
    Homecraft Classes

168 Victoria Sandwich

169 *Victoria Sandwich Recipe*

172 Vegetable Cakes

173 Drizzle Cakes

174 *The Cake Baker*

175 *Sarahjane's Beetroot
    Cake Recipe*

176 Fruit Cakes

177 Children's Bakes

178 Scones

179 Shortbread

180 *The Shortbread Baker*

181 *Joy's Shortbread Recipe*

182 Lemon Tart

183 *Lemon Tart Recipe*

184 Quiche

185 *Quiche Lorraine Recipe*

186 Sausage Rolls

187 Bread

188 *The Fancy Bread Baker*

189 *Richard's Tear'n'Share
    Bread Recipe*

190 Chutneys & Sweet Pickles

191 *Piccalilli Recipe*

192 *The Judge of Preserves*

196 Marmalade

197 *Vivien's Marmalade Recipe*

198 Jams

199 Jellies

200 *The Jam Maker*

201 *Debbie's Apricot &
    Passion Fruit Jam Recipe*

202 Curds

203 Fruit Liqueurs

204 *The Show Organiser
    & Curd Maker*

205 *Sue's Lemon Curd Recipe*

NUTS & BOLTS

210 Organising a Show

212 Writing a Show Schedule

214 Show Rules

218 Glossary

220 Index &
    Acknowledgements

Foreword by Alan Titchmarsh

There is nothing else like it on earth. The Olympic Games simply cannot compete with the tensions, the pressure, the rivalry and the subsequent triumphs encountered at the Great British village show. It is here, under canvas or the rafters of the village hall, that bakers of Victoria sponges, makers of jams and marmalades and growers of gooseberries and unreasonably long parsnips compete for local notoriety and the respect accorded to those who are best in their field – or their kitchen.

But there is more to the village show than rivalry. Yes, emotions come to the fore, and occasionally friendships might be strained, but above all the growing of fruits and vegetables, flowers and plants, and the creation of cakes and biscuits, pastries and preserves, is a celebration of what it means to be part of a community that values the skills necessary to grow things and cook things. And that, for me, is why the village show is unique: it not only celebrates bakers and cooks (and I speak as someone who loves eating), but it also applauds those who cultivate their ingredients. It is a marriage of skills and a moment in the year when those in the community can get together and demonstrate their craftsmanship and know-how.

In a world that seems to be increasingly distanced from rural life, a village show demonstrates in its own simple way the continued loyalty to the traditions of the countryside – the cultivation of food and the creativity involved in bringing it to the table.

We can all marvel at the largest cabbage and pumpkin, and at a lattice tart or Simnel cake of such perfection that we know it would be beyond our own powers to create; but we are enriched by the fact that there are those among us who are possessed of these skills. The great thing about this book is the generosity shown by those who are at the top of their game when it comes to making jams and cakes and growing fruit and vegetables that regularly win the scarlet rosette. Within these pages you will find their advice – a mixture of sound common sense and insider know-how – that might just persuade you to take part in your own local show. It is all too easy to stand on the sidelines and admire, and to be intimidated by what we see laid out on doilies and shining plates. But no-one to my knowledge has ever been mocked for entering a village show. Far from it; you will be welcomed with open arms. You'll also find that competing broadens your skills, improves your growing and baking techniques and helps bind a local community together.

Taking part in a village show is the best demonstration that we care about our gardens, our countryside and our rural communities. And if your sponge falls flat and your tomatoes don't make the grade, don't worry; you can still take them home and eat them. Good luck!

Alan Titchmarsh MBE VMH DL
Vice-President, Royal Horticultural Society

Growing for the Show
MATTHEW BIGGS

At any village show you'll see super-keen gardeners with entries in every category alongside first-timers tentatively offering up one or two exhibits. All are welcome, whatever their experience.

Plan early to boost your chance of winning. Do think carefully about the classes you want to enter and work out how much time you can realistically spend on nurturing your crop – beware of being overambitious. Unless you have your own favourites already, it's generally a good idea to grow cultivars that are proven show-winners.

Attention to detail can be key. Give your plants what they need to thrive, then keep a constant eye on them, and stay one step ahead of pests and diseases. Experienced growers prepare raised beds for their root veg, devise secret compost recipes to bring on their crops, and tap into traditional techniques in an effort to receive that single extra winning point ahead of their rivals.

Prepare to catch the exhibiting bug. Talk to any village show veteran and they will say the same thing: "After a few wins, I became hooked". It's the buzz of success – not to mention local bragging rights – that will bring you back year after year.

Cooking for the Show
THANE PRINCE

A wonderful celebration of amateur talent, the village show unearths all the local star bakers and jam-makers. Jewel-coloured jellies and chutneys fill the show tables, along with tempting cakes and bakes.

Decide where your talents lie. Are you an ace jam-maker, a perfect cake-baker, or an expert on the slow arts of sloe gin? With so many classes to choose from, it can be tempting to enter a whole host of categories, but you may well find success comes with an honest examination of your skills.

Always check the entry criteria thoroughly. Nobody wants to bake a perfect Victoria sponge and then fill it with buttercream when only jam is permitted; or to discover at the show bench that their jars of chutney are too small or incorrectly labelled. It is on such simple details that victory can depend.

Enjoy the day. Once your cherished exhibit is staged, it's out of your hands. All you can do is sit back, wait for the judges' verdict, and hope to experience the surge of pride that comes from finding a "Best in Show" certificate beside your exhibit.

GROW TO SHOW
FRUIT

Learn how to stage everything from dainty summer berries to autumnal tree fruits, and discover which apple and pear cultivars are just perfect for showing.

Tips & Tricks
FRUIT CLASSES

The colourful array of fresh fruit laid out on the show bench can look utterly mouthwatering. Visitors never see the effort put in to growing, but they will notice if fruit has been poorly presented or damaged in transit – as will the judges...

Individual Fruit Classes

PREPARATION Harvest fruits as near to show time as possible, taking care not to damage them. Avoid any malformed specimens. Pick more than is required, in case any is damaged

A Class of their Own

Most shows will provide a general class for any fruits not listed in their own individual fruit class. These general classes can receive a wide variety of entries, with potentially different point values. For fairness, larger shows may provide two "any other" classes: one for lower-value entries, and one for higher-value ones. Smaller shows may also make allowances for the difficulty and standard of exhibit when judging a wide variety of entries, to ensure that a poor-quality exhibit of high value does not automatically beat a top-quality exhibit of low value. If an exhibitor grows exotic or unusual fruits, a general class is often the place to enter them, but they must still be good quality.

during transit. In most cases, specimens should be ripe and their stalks should be retained, but there are exceptions and you should check the individual entry to confirm; for example, quince may be presented when not fully ripe. If specimens are supposed to be presented without stalks (e.g. apricots), take care not to split the skin in the stalk cavities of the fruit. Handle specimens gently, by their stalks (if attached), to avoid spoiling the natural bloom of the fruit. Under no circumstances should fruit be polished.

TRANSIT Pack fruits carefully, in a cardboard box or other suitable container. Avoid packing too many into a single box, in case the fruits bruise or squash each other. Take particular care with soft fruits, bearing in mind that they may be damaged by their own weight. Use tissue or newspaper to line the box and the spaces between the fruits, if necessary. Store fruits in a cool place before transporting to the show.

PRESENTATION Check all specimens as you unpack them, rejecting any that are damaged. Aim for a neat, attractive, symmetrical staging, following the presentation advice provided in individual entries. As a general rule, apples and similar-shaped fruits should be staged with the stalk end facing down.

Mixed Collections and Trug Exhibits

PREPARATION Check the schedule to confirm the quantities and requirements of any mixed collection and/or trug classes offered. Mixed collection classes tend to require a set number of specimens, while trug classes usually specify a minimum number of kinds of fruit to be displayed (e.g. four kinds of gooseberries).

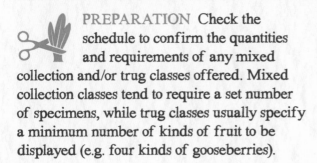 **TRANSIT** For both mixed and trug classes, pack fruits as for individual class exhibits (see left). Assemble trug and basket displays directly on the show bench, in the allotted space; do not transport specimens to the show in the trug, or attempt to lift a filled decorative trug by its handle, as this may damage both the trug and the fruit within.

PRESENTATION Stage mixed collections neatly, as you would if entering each fruit type in its individual class. Fill trugs and baskets with the specimens in an artful arrangement; in these classes, presentation is just as important as the quality and variety of produce, with up to 20 points available in each case. Organisers may provide trug exhibitors with a set space on the show bench (e.g. 90cm x 90cm/3ft x 3ft), which the display must not exceed.

Keep on Truggin'

Mixed collection and trug classes have grown in popularity at shows in recent years. Requiring a mastery of varied cultivation regimes, these classes demonstrate the all-round ability of the exhibitor. They can range from simple exhibits of three or four different cultivars of the same fruit, to grand trugs filled with a cornucopia of many different, perfectly ripe produce. A well-filled, colourful trug makes an attractive display, and can easily become the star exhibit of the show bench.

Staging the Exhibits

It's an early start for exhibitors, who need to safely transport their entries to the show and neatly present them on the show bench.

The tent is a flurry of activity on the morning of the show: the show tables are set out, the bunting is hung, and the exhibitors arrive to stage their entries.

Super-sized contenders are often weighed and staged the day before, when heavy lifters are on hand with wheelbarrows to help move the entries. In extreme cases, a forklift truck may need to be called in.

If you're new to showing, look around at the other exhibitors and see what they are doing. Should you be trimming your dahlias to the same height? Could your tomatoes do with another clean before staging?

Strawberries

Nothing is more redolent of British summertime than a bowl of perfectly ripe strawberries. Whether entering regular or alpine cultivars, your berries must be red from hull to tip.

JUDGES SCORE HIGHLY
Uniformly large, ripe fruits, bright red in colour, and in good condition with no blemishes. Calyces (hulls) should be fresh and stalks intact.

DEFECTS TO AVOID
Small, dull, un- or overripe berries, in poor condition or "hard nosed", lacking stalks and calyces.

POINTS AVAILABLE

REGULAR
Condition	4
Uniformity	4
Size	4
Colour	4
TOTAL	16

ALPINE
Condition	2
Uniformity	2
Size	2
Colour	2
TOTAL	8

PREP Cut berries with scissors, leaving a portion of stalk attached. Pick more than you need in case any are damaged in transit. Select fully ripe berries and avoid any that are misshapen, or have green or hard noses, which indicate poor fertilisation of the flowers.

PRESENT Lay berries in neat lines or in a circle, with all stalks pointing the same way. Handle by the stalks to avoid bruising fruits.

JUDGING NOTES For regular strawberries, competitors will usually need to show 15 specimens, or a minimum of 8 at smaller shows. For alpine strawberries, 25 fruits are normally required, or a minimum of 13 at smaller shows. Both types are judged by the same criteria. If there is no class for alpines they may be entered into a miscellaneous class.

Small and Sweet
Alpine strawberries are not much larger than peas, but what they lack in size, they make up for in an exquisitely sweet taste.

Pick strawberries with their stalks and hulls intact

Judges look for a fresh, bright red colour

Fruits should be uniformly large and ripened all over

Raspberries

Traditionally a midsummer treat, raspberries are now exhibited through to October thanks to new autumn-fruiting cultivars. Choose a cultivar that will mature in time for your local show.

PREP Cut ripe berries with scissors, leaving the calyx (hull) and a portion of stalk attached. Handle with care (by the stalk, if possible) and pick more than you need in case of damage.

PRESENT Arrange the berries neatly and symmetrically. They look attractive placed in lines, stalks pointing in the same direction, or can be arranged in a circle around the rim of the plate, stalks pointing out.

JUDGING NOTES You will normally be asked to show 20 specimens, or at least 10 at smaller shows. Most shows will offer a general "soft fruits" category if a specific category for raspberries is not listed in the schedule.

JUDGES SCORE HIGHLY
Large, ripe berries of good colour, blemish-free, with fresh calyces (hulls) and stalks attached.

DEFECTS TO AVOID
Small, un- or overripe berries, with a dull colour, or blemished by insects or poor fertilisation. Lack of stalks.

POINTS AVAILABLE

Condition	4
Uniformity	2
Size	3
Colour	3
TOTAL	12

TOP TIP
Birds and squirrels can strip a whole crop of berries, usually just a few days before the show! As soon as fruits start to form, cover the plants with protective netting and check carefully for gaps, as the hungry pests will certainly find any...

Place berries carefully and avoid leaving stains

Fresh leaves can create an attractive foil to the berries

Blackberries & Hybrids

In the wild, hundreds of different microspecies of blackberry grow with reckless abandon, but it is fruits of the well-behaved garden types, with their uniform taste and appearance, that grace the tables at village shows.

JUDGES SCORE HIGHLY
Fully ripened fruits of uniform colour, shape, and size, in a good, blemish-free condition, with fresh calyces (hulls) and stalks.

DEFECTS TO AVOID
Small, dull, un- or overripe berries and those with insect damage or defects caused by poor fertilisation. Lack of stalks.

POINTS AVAILABLE

Condition	4
Uniformity	3
Size	3
Colour	2
TOTAL	12

PREP Use scissors to select large, ripe berries with calyces (hulls) and a length of stalk attached; take care handling the stems of thorny cultivars. Choose berries that have fertilised completely, so that all "drupelets" (the many individual globules that make up each fruit) are fully formed. Ripe berries are fragile and can easily be damaged between picking and presenting, so pick more berries than you need and protect them with cotton wool or tissue paper.

PRESENT Lay the berries out in lines with stalks pointing the same way; in a circle with stalks pointing out; in a neat combination of the two; or even in full sprigs.

JUDGING NOTES Competitors are normally asked to show 15–20 berries, or a minimum of 10 at smaller shows. Be sure to confirm the quantity of specimens required for any particular cultivar: if the schedule is unclear, try contacting the show organisers directly.

Fantastic Discovery

The larger-than-life blackberry known as 'Fantasia' is a village show favourite and it was discovered quite by chance in deepest suburbia. A certain Mr Keates found an unusually vigorous blackberry seedling growing on his allotment wall in Surbiton, Surrey. He realised its commercial potential and, over time, the happy mutation became available to buy. Although viciously spiny, 'Fantasia' has the redeeming feature of producing very tasty and very large fruit – the size of a 50p piece – in abundance.

Blackberries
(Long Cultivar)

*Hulls and stalks
should be attached*

*Berries should
have a natural,
bright colour*

*A strig of perfect
berries makes an
impressive entry*

Blackberries
on a Strig

KNOW YOUR HYBRID BERRIES

"Hybrid" berries are mainly the result of cross-breeding blackberries with other species in the *Rubus* genus. For show purposes, they are judged by the same criteria as blackberries and scored the same.

Boysenberry

Developed in the US in the 1920s by breeder Rudolph Boysen, they are slightly larger than blackberries and have a more tart flavour, perfect for jams and pies. Delicate and prone to bleeding: handle with care!

Loganberry

An accidental raspberry-blackberry cross discovered by US breeder James Harvey Logan in 1881, loganberries are similar in size to blackberries and taste like sharp raspberries.

Tayberry

Developed in the Dundee area and named after the River Tay, tayberries are larger and sweeter than loganberries.

Wineberry

Not a hybrid at all but an Asian species of raspberry (*Rubus phoenicolasius*), wineberries look like red blackberries and have an intensely sweet flavour.

Currants

Available in black, white, red, and now a pretty pink too, currants are perfectly suited to our climate, preferring cool, moist conditions. Net currants against birds or they will happily strip the bush and leave none for the show bench.

JUDGES SCORE HIGHLY
Strigs with a full complement of large, ripe, uniform berries.

DEFECTS TO AVOID
Berries missing on strigs, shrivelled stalks, unevenly ripened berries.

POINTS AVAILABLE
Condition	3
Uniformity	3
Size	3
Colour	3
TOTAL	12

PREP Use scissors to harvest currants with the strigs intact, choosing the longest strigs with the largest berries (strig is the technical term for the fruiting flower stalk). Handle carefully, holding by the stalk end of the strig, to avoid knocking off or damaging berries.

PRESENT Currants may be neatly spread across the centre of a plate or formed into a ring, especially if the cultivar produces naturally short strigs (see Judging Notes), but ideally strigs should be laid roughly parallel with the stalk ends pointing out.

JUDGING NOTES Entrants are normally asked to exhibit a dish of strigs weighing 200–250g (7–9oz), but smaller shows may specify a certain quantity of individual strigs. Some modern cultivars do not produce long strigs and in such cases it is permissible to show a small portion of woody stem.

Currants now appear in shows from June to September thanks to new early and late cultivars.

Blackcurrants for Britain
Currants contain up to four times more vitamin C than oranges and have traditionally been used for centuries to treat colds and sore throats. When oranges became scarce during the Second World War, the entire British blackcurrant crop was devoted to the production of syrup and distributed to children to maintain their good health. Our national love of blackcurrant squash has held pretty firm ever since!

Redcurrants

Red currants
should be jewel-like,
shiny and bright

Pink cultivars are sweet
enough to be eaten
straight from the bush

Pink
Currants

Points will be deducted
for strigs on which the
berries have ripened
unevenly

White
Currants

Gooseberries

Competitions to grow the heaviest gooseberry are traditional in northern England, where you'll find berries the size of duck eggs. In the less weight-obsessed world of the ordinary village show, large berries are still preferred.

JUDGES SCORE HIGHLY
Large, uniform, unblemished fruits of good colour for the cultivar, and with stalks intact.

DEFECTS TO AVOID
Fruits of uneven size and any that are overripe, split, or diseased.

POINTS AVAILABLE
Condition	4
Uniformity	3
Size	3
Colour	2
TOTAL	12

PREP Leave a short length of stalk intact when picking berries. Fruits may be picked ripe or unripe, according to the cultivar type and the time of year (see below).

PRESENT Gooseberries look best arranged in a circle around the edge of a plate with the stalks pointing inwards and the "tails" (the remnants of the flowers) pointing out.

JUDGING NOTES For a single-dish exhibit, 20 fruits are normally required, or a minimum of 10 at smaller shows. Uniquely amongst fruit categories, gooseberries may be submitted for competition when still relatively unripe, provided the show being entered takes place early in the fruiting season for the cultivar being shown. This is because most gooseberry cultivars are classified as both "culinary" and "dessert", which means they can be enjoyed unripe if cooked with additional sugar or eaten straight off the bush later in the season when fully ripe. Dessert-only cultivars should be shown fully ripe at competition.

Berry Basket
Show schedules may include a collection competition for three or more gooseberry cultivars. Unusual yellow and white varieties here sit alongside the traditional green and red.

EXHIBITOR'S NUMBER

Fruits should have
a good, even colour
for the cultivar

Green
Gooseberries

Grow a Green Giant

Growing giant gooseberries requires
a great deal of feeding and watering,
a constant lookout for pests and
diseases, and above all a willingness
to sacrifice your crop, as the bushes
must be stripped of most of their
berries so that those left behind
can expand to giant proportions.
Champion exhibitors keep their
methods top secret, but rumour has
it they will thin fruits out to a one
per branch or even a mere one fruit
per bush, thereby channelling all
the energies of the plant into
swelling that sole remaining
berry. The heaviest
gooseberries can grow
up to five times
the size of an
average berry.

Average berry

Giant berry

Choose large,
plump berries that
match as closely
as possible in size

Arrange berries with
tails facing out and
stalks turned inwards

Red
Gooseberries

Some cultivars
naturally produce
smaller berries

Red
Gooseberries

Cherries

Whether you grow sweet cherries (*Prunus avium*) to eat raw or sour cherries (*Prunus cerasus*) for making pies and jam, be sure to enter them in the correct class to avoid any risk of disqualification.

JUDGES SCORE HIGHLY
Large, ripe, unblemished fruits of a brilliant colour, with a natural bloom and unshrivelled stalks.

DEFECTS TO AVOID
Small, un- or overripe fruits, or those with a dull, split, or otherwise blemished surface. Absence of stalks or shrivelled stalks.

POINTS AVAILABLE

SOUR
Condition	3
Uniformity	3
Size	3
Colour	3
TOTAL	12

SWEET
Condition	4
Uniformity	4
Size	4
Colour	4
TOTAL	16

TOP TIPS
Sweet cherries need a sheltered position and plenty of sunshine in order to develop their sweetness. Sour cherries don't need to ripen to the same level of sweetness and are therefore hardier and better suited to cooler, north-facing gardens.

PREP Choose large specimens of top-notch colour for their cultivar. Pick the fruit carefully, leaving the stalks intact and the bloom undisturbed.

PRESENT Cherries can be laid out in lines across the plate or they can be presented in a circle, if preferred. Whichever you choose, make sure the stalks are arranged neatly: either all going in the same direction if displayed in lines or all pointing outwards if in a circle.

JUDGING NOTES Competitors will normally need to present 20 specimens, or a minimum of 10 at smaller shows. Sweet cherries are harder to grow in the UK and therefore garner more points than sour cherries (see Top Tips).

Aim for a neat, symmetrical layout

stalks should remain attached

Judges prefer fruits with a brilliant colour

Apricots

Apricots hail from sunny Armenia, and although this well-travelled fruit prefers hot, dry weather, apricots are appearing more frequently at shows thanks to new cultivars suited to UK conditions.

PREP Use scissors to remove ripe fruits from the tree. Unlike many tree fruit, apricots are shown without the stalks attached. Take care not to split the skin when removing the stalks.

PRESENT Lay apricots out in lines or in a circular arrangement. Apricots have a very delicate skin that marks easily, so handle them carefully to prevent bruising. Avoid removing any of the "bloom" on the surface of the fruits; this is a natural, powdery surface coating, which the judges will be looking out for.

JUDGING NOTES Competitors are normally asked to show 6 fruits, or a minimum of 3 fruits at smaller shows.

JUDGES SCORE HIGHLY
Large, uniform fruits that are highly coloured with a clear skin, no blemishes, and a good bloom.

DEFECTS TO AVOID
Small or misshapen fruits and ones that are un- or overripe, have any blemishes, show signs of decay, splitting, or bruising, or still have their stalks attached.

POINTS AVAILABLE

Condition	4
Uniformity	4
Size	4
Colour	4
TOTAL	16

Cool-climate Cultivars

Apricots have been grown outdoors in Britain since Henry VIII's gardener brought trees back from Italy. For years 'Moorpark', bred at the Moor Park estate in Hertfordshire in the 18th century, has been the one to grow. Recently it has been joined by 'Flavorcot', 'Tomcot', and patio cultivar 'Aprigold', bred to produce quality fruit in our cooler climate. Protect apricot's early blossoms against frost with horticultural fleece, but roll up the cover during the day to allow insects in to pollinate.

Choose perfectly ripe fruits

Skin should be clear and highly coloured, with a good bloom

Size should be above average for the cultivar, but not overly large

Fruits should be free of stalks

Plums

Plums are a good tree fruit for show novices to grow, as they are easy to prune and pick. Sweet dessert plums and the tarter cooking cultivars should be entered into separate classes.

JUDGES SCORE HIGHLY

Large, ripe fruits, of good colour, with stalks and bloom intact. Cooking plums should be firm, with perfect bloom.

DEFECTS TO AVOID

Small, un- or overripe fruits of poor colour, imperfect bloom, or lacking stalks. Slight shrivelling in gages is not a defect, as it is a sign of ripeness.

POINTS AVAILABLE

DESSERT

Condition	5
Uniformity	4
Size	3
Colour	4
TOTAL	16

COOKING

Condition	4
Uniformity	3
Size	4
Colour	3
TOTAL	14

PREP Cut plums from the tree as near to show time as possible, leaving the stalks and skins intact. Avoid damaging the bloom by handling as little as possible and holding them by their stalks. Cooking plums should be ripe, but not too soft. Dessert plums should be soft and fully ripe.

PRESENT Stage on their sides on plates, either in a circle with stalks facing in, or in symmetrical lines with stalks pointing the same way.

JUDGING NOTES Schedules normally require competitors to enter 9 plums, or a minimum of 5 plums at smaller shows. Dessert and cooking plums are judged in different classes, so make sure you enter the correct class to avoid disqualification. If you are not sure how your plum should be classed, a comprehensive list of dessert and cooking cultivars is available from the RHS. Gages are categorised as dessert plums, as are mirabelles or cherry plums.

TOP TIPS

Plum tree blossom, which appears between March and April, needs frost protection. Prevent birds from eating the fruit buds by growing your trees as a "fan" against a wall, then covering with netting.

Points are awarded for specimens retaining their natural bloom

Plums in a circle should have stalks facing inwards

Dessert plums should be soft and fully ripe

Judges are likely to dock points if stalks are missing

Damsons should be ripe but can be relatively firm to the touch

Try to preserve the delicate bloom on the skin

The Road from Damascus?
"Damson" derives from "damascene" and it was once thought they were brought back to Britain from Damascus by Crusaders. Current thinking suggests damsons are a cross between native bullaces and wild sloes.

Damsons & Bullaces

Though types of cooking plum, these fruits are usually shown in a separate class. Bullaces are smaller and rounder than damsons and have a strong claim to being Britain's only native plum.

PREP Like cooking plums, fruits should be ripe when picked but need not be soft. Pick them carefully so as not to tear the skin around the stalk, and handle as little as possible to avoid wiping off the bloom, which is a highly sought after feature of these plum types. Pack them in tissue paper to preserve the bloom in transit.

PRESENT Stage on a plate, either in a circle or in symmetrical lines.

JUDGING NOTES Competitors are normally required to show 9 specimens, or a minimum of 5 at smaller shows. If there is no separate class for damsons and bullaces they may be entered into the class for cooking plums.

JUDGES SCORE HIGHLY
Large, ripe but firm fruits, of good colour, with perfect bloom.

DEFECTS TO AVOID
Small, unripe fruits or fruits so ripe as to be soft; fruits showing poor colour and with imperfect bloom.

POINTS AVAILABLE

Condition	2
Uniformity	2
Size	2
Colour	2
TOTAL	8

Gieser Wild
Culinary /

Figs

Fig trees are surprisingly easy to grow in British gardens, but getting them to produce sweet, succulent, show-worthy fruits takes skill and understanding.

JUDGES SCORE HIGHLY
Uniform, large, fresh, fully ripe fruits, of good colour, with their natural bloom and stalks intact.

DEFECTS TO AVOID
Small, unripe fruits of poor colour or with imperfect bloom, and any without stalks.

POINTS AVAILABLE

Condition	5
Uniformity	3
Size	5
Colour	3
TOTAL	16

PREP Cut the fruit down with secateurs or sharp scissors, leaving their stalks intact. Figs stop ripening once cut, so only cut when large, well-coloured, soft to the touch, or cracked at the base. Handle very gently, by their stalks if possible, to preserve their natural bloom.

PRESENT Stage the fruits in a circle on a plate; round figs look best sat with stalks uppermost and long-bodied figs are best placed flat on the plate with the stalks in the centre.

JUDGING NOTES Entrants are normally asked to exhibit 5 figs, or a minimum of 3 figs at smaller shows. Figs may still be exhibited with skins that are starting to split, especially at the base, and with some drops of nectar showing, as these are signs of ripeness.

The fruits should be uniformly large and ripe

Make sure the natural bloom is unspoilt

Keep the stalks attached

Long Gestation
Grown outdoors in Britain, figs will typically crop once a year, and growing fully ripe fruits is a two-year process. At summer's end, trees bear embryo fruitlets the size of a pea. Protect these little fruits over winter with horticultural fleece, and they should ripen the next year.

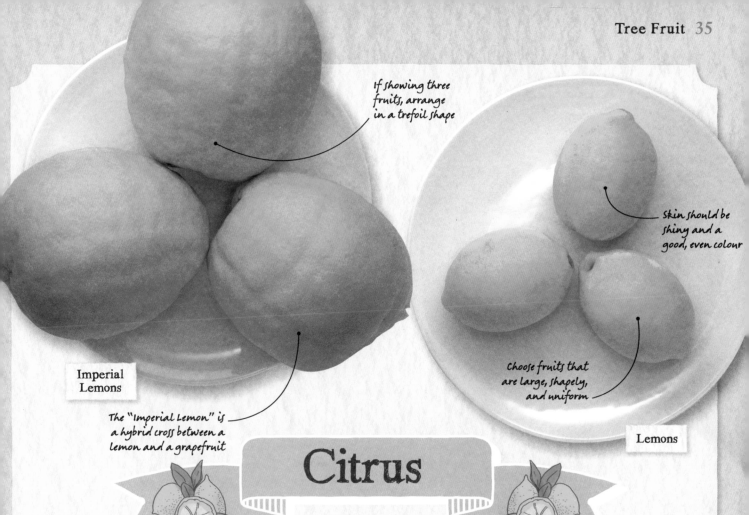

If showing three fruits, arrange in a trefoil shape

Skin should be shiny and a good, even colour

Imperial Lemons

choose fruits that are large, shapely, and uniform

The "Imperial Lemon" is a hybrid cross between a lemon and a grapefruit

Lemons

Citrus

In Britain, citrus plants are best grown in pots, to soak up the sun outdoors in summer and then be moved under glass for winter. Ripe fruits can be left on the plant and harvested on show day.

PREP Pick the fruits without any attached stalk and stem, being careful not to split the skin in the stalk cavity. Do this when they are fully coloured and seem to have stopped growing, and as close to the day of the show as possible.

PRESENT Arrange in a trefoil shape in the centre of the plate, with stalk cavities facing inwards.

JUDGING NOTES Entrants are normally asked to show 3 citrus fruits (excluding kumquats and calamondin oranges), or a minimum of 2 fruits at smaller shows. For calamondins and kumquats, it is normally 9 fruits that are required for an entry, or a minimum of 5 at smaller shows.

JUDGES SCORE HIGHLY
Uniformly large, shapely, ripe fruits of a good, even colour that is natural to the cultivar; bright, shiny, unblemished skins, and no stalks.

DEFECTS TO AVOID
Small, misshapen, un- or overripe fruits of a dull or uneven colour, or any with blemishes or their stalks still attached.

POINTS AVAILABLE

Condition	5
Uniformity	5
Size	4
Colour	4
TOTAL	18

Apples

Apple exhibitors of a pagan mindset have been known to honour their trees by drinking a "wassail" of mulled cider. You needn't show such devotion to your apples, but do make certain of their identity and enter them into the right class.

JUDGES SCORE HIGHLY
Shapely, uniform fruits with clear, unblemished skins of good colour characteristic of the cultivar. Dessert apples should be in the correct size range; cooking apples should be large and solid.

DEFECTS TO AVOID
Misshapen, overripe, or soft fruits that have damaged eyes, lack stalks, or have any blemish, including evidence of a physiological disorder such as bitter pit (dry, dark sunken spots) or glassiness (water retention in the flesh). Dessert apples should be neither too small nor too large, while cooking apples should not be small.

POINTS AVAILABLE

DESSERT	
Condition	6
Uniformity	6
Suitability of size	4
Colour	4
TOTAL	20

COOKING	
Condition	6
Uniformity	6
Size	6
TOTAL	18

PREP Cut with secateurs to leave the stalk intact and handle fruits as little as possible by their stalks. Apples are one of only a few fruits (others include gooseberries and medlars) that may be exhibited either ripe or unripe.

PRESENT Stage on plates with the stalk ends facing down, either in a ring or with one fruit in the centre, which can be raised.

JUDGING NOTES Entrants are normally asked to submit 6 specimens in both cooking and dessert classes, or a minimum of 3 specimens at smaller shows.

Make sure you enter your apples in the correct class or they may be disqualified. If you are unsure what class your apples should be entered into, you can check the list of dessert, cooking, and dual-purpose cultivars available from the RHS. Dessert apples: a diameter of 60–80mm (2^1/$_3$–3^1/$_8$in) is preferred, but judges will make allowance for the fact that some cultivars are inherently small, whereas others are naturally large. Examples of inherently small dessert apples are: 'Margil', 'Pitmaston Pine Apple', 'Sunset', and 'Winston'. Examples of inherently large dessert apples

Cooking apples should be large and uniform in size

Cooking Apples

Cooking
Apples

*Choose apples
with clear,
blemish-free skin*

include: 'Belle de Boskoop',
'Blenheim Orange', 'Charles
Ross', 'Gascoyne's Scarlet',
'Jonagold', 'Jupiter', 'Reinette
du Canada', 'Rival', and
'Wealthy'. Cooking apples: in
a cooking apple class, provided
two entries are equal in all other
respects, the entry with the larger
specimens relative to average
cultivar size will be preferred
by the judges. Dual-purpose
apples: a few apple cultivars are
considered suitable for exhibiting
as either dessert or cooking apples,
according to size, and may be
shown in one or other class, but not
both. Ideally, when a dual-purpose
cultivar is shown as a dessert
apple, fruits should not exceed
75mm (3in) in diameter. When
shown as cooking apples, fruits
should comfortably exceed the
minimum size of 80mm (3$\frac{1}{8}$in).
Over-size or under-size would not
disqualify an exhibit. In collection
classes, the same cultivar cannot be
shown as both dessert and cooking.

*Fruits should show the
typical colouring of
their cultivar*

Dessert
Apples

Apple-tree, Apple-tree, Bear Good Fruit

To grow apples you'll need to buy
at least two trees, as most need a
pollination partner – a variety with
compatible pollen that flowers at the
same time – to ensure good fruiting.
Consider also when the last frosts
tend to occur in your area and choose
varieties that flower after this time,
as a late frost can kill off blossoms
(and your crop) in one fell swoop.

A World of Apples

An autumn show is a great place to discover the fabulous range of apples still grown in Britain. If you are inspired to start your own orchard, seek out unusual and heritage cultivars and help keep diversity alive in the world of apples – you can often try before you buy. Here's a selection of beauties to get you started.

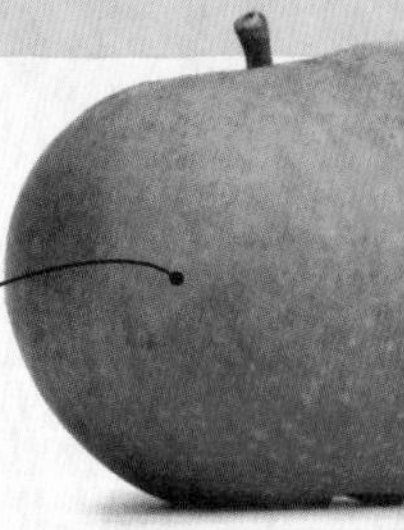

Discovered in 1740 near Blenheim in Oxfordshire

'Blenheim Orange'

'Red Devil'

'Duke of Devonshire'

'Jonagold'

'Egremont Russet'

'Rubinstep'

'Scrumptious'

'Laxton's Superb'

'Peasgood's Nonsuch'

'Kidd's Orange Red'

A modern cultivar developed at East Malling Research Station in Kent

'Saturn'

'Limelight'

Unidentified apples can be exhibited as 'Unknown Seedling'

'Unknown Seedling'

'Greensleeves'

'Herefordshire Russet'

A modern, stronger-tasting rival to the 'Egremont Russet'

'Red Jonaprince'

Stand fast root, bear well top Pray send us a howling good crop. Every twig, apples big. Every bough, apples now.

~ 19th-century apple wassail rhyme from Sussex

Intensely flavoured apple raised in Gloucestershire in the early 18th century

'Ashmead's Kernel'

'Cox's Orange Pippin'

'Spartan'

'Howgate Wonder'

'Lord Derby'

'Lord Hindlip'

Named after a Croydon vicar and former Secretary of the RHS

'Reverend W. Wilks'

'Golden Delicious'

Cox-style cultivar raised by the Laxton Brothers in the early 1900s

'Laxton's Fortune'

'Charles Ross'

'Bramley's Seedling'

The RHS Judge

My name is Colin Spires, and I've been judging shows at all levels for over forty years.

MY FIRST SHOW

I judge all sorts now – fruit, vegetables, and flowers – but I started out as a dahlia judge. For my first judging engagement, I was kindly invited to take part in a local show. What I hadn't realised was that this show was one of the top dahlia shows of its kind, all national-standard exhibitors. It was also a publicly judged event: I had to stand in front of an audience and explain my judging to them as I went along.

Well, I looked at one of the first exhibits and said, "this is wrongly named". It was a 'Nina Chester', but had been labelled 'Banker'. Next thing I know, a voice pipes up, saying "don't be stupid, that's Banker!" It didn't make a difference to my judging, but I told the man the flower was definitely named incorrectly.

Further along the bench, I came across another flower – labelled 'Nina Chester'. "Ah," I called out to the heckler, "you've got your labels mixed up!" Everyone chuckled; they knew I was right. It was a fun way to start my judging career.

> ## "I've seen shows so warm that the flowers are wilting as you're looking at them. You've got to take that into account."

Classing Apples

The Horticultural Show Handbook is the bible for RHS judges and Colin provided revisions for the latest edition, including an update to the list of apple cultivars.

EXOTIC IDEAS

Schedules often offer an "open" class for any fruit and vegetables that don't have a class of their own. Sometimes, competitors will try exhibiting something more unusual in this kind of class, submitting exhibits like melons and kiwi fruit.

In my experience, unusual or exotic exhibits tend to be quite poor, at least in comparison to tried-and-tested show classics. People think that, just because it's exotic, it should win. But that isn't true: as a judge, you always look at the quality of the exhibit, not how unusual it is.

Another, more practical problem with exotic fruit and vegetables is that, at least at the level of a small village show, the judges may never have grown, or even judged, the produce that's in front of them. That's not to say they can't judge it – they'll use their common sense. But it's something competitors have to bear in mind if they want to enter a honeydew melon into a show that mostly sees currants: no matter how unusual it is, it still clearly needs to be a show-quality specimen.

JUDGING STANDARDS

These days I probably judge about 30-odd shows a year, ranging from local village fêtes up to the Chelsea Flower Show. As you can imagine, there's a big difference in standard between these types of shows.

Now and then, I go to an amateur competition and find someone that really stands out: a dahlia expert, say, showing brilliant dahlias, while everyone else is just an ordinary gardener having a go with one or two flowers. That doesn't mean I'll suddenly raise the bar to an expert level and start judging the amateur exhibits more harshly. I do what the RHS advocates: before I start judging, I look around all the tables (even the ones I'm not judging) to see the general quality of exhibits, and use that average standard as a starting point for my marking.

Flower Power
A bad example of judging at his local show prompted Colin to take the National Dahlia Society exam and become a judge himself.

"Just because something is exotic doesn't mean it's guaranteed to do well in a show – it still needs to be top quality."

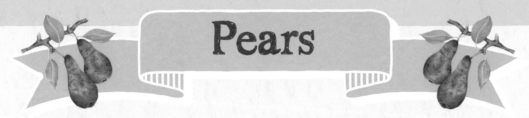

Pears

Pears are less common than apples in both gardens and in competition, perhaps because they are more pernickety, being at greater risk of frost and wind damage and needing more warmth and sunshine to crop.

JUDGES SCORE HIGHLY
Large (for the cultivar), uniform, shapely fruits, with undamaged eyes, their stalks intact, and clear, unblemished skins of a colour characteristic of the cultivar.

DEFECTS TO AVOID
Fruits that are small, misshapen, overripe, or shrivelled; fruits with damaged eyes, blemishes, no stalks, and any that are not well-coloured. In addition, cooking pears should not be overripe or soft.

POINTS AVAILABLE

DESSERT
Condition	6
Uniformity	6
Size	4
Colour	4
TOTAL	20

COOKING
Condition	6
Uniformity	6
Size	6
TOTAL	18

PREP Pick pears as near to show time as possible. Use secateurs or sharp scissors to cut fruits from the tree, leaving the stalk intact. Handle fruits as little as possible, by their stalks. Pears can be shown either ripe or unripe, but judges tend to prefer good examples of ripe cultivars in season rather than larger or more showy cultivars that have been picked prematurely.

PRESENT Most pears are best staged flat on the plate in a neat arrangement with stalks towards the centre. Some of the rounder cultivars, however, may be staged like apples (see pp.36–37).

JUDGING NOTES Entrants are normally asked to show 6 fruits, or a minimum of 3 at smaller shows. Be sure to enter your pears into the correct class or they will be disqualified. The following are classified as cooking pears: 'Bellissime d'Hiver', 'Beurré Clairgeau', 'Black Worcester', 'Catillac', 'Uvedale's St Germain', and 'Vicar of Winkfield'.

Choose pears that are large for the cultivar

The skin should be clear and unblemished

keep the stalks intact

A World of Pears

The choice of pears in supermarkets is even more restricted than for apples, and many visitors to autumn shows are amazed to find so many different colours, shapes, and sizes on the show bench.

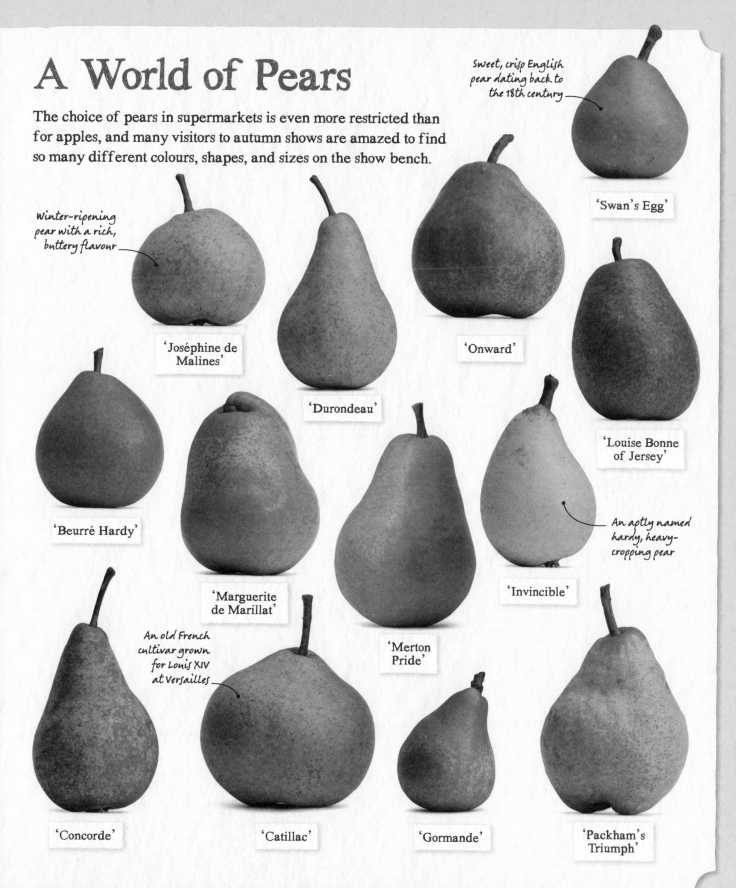

Sweet, crisp English pear dating back to the 18th century

'Swan's Egg'

Winter-ripening pear with a rich, buttery flavour

'Joséphine de Malines'

'Onward'

'Durondeau'

'Louise Bonne of Jersey'

'Beurré Hardy'

An aptly named hardy, heavy-cropping pear

'Invincible'

'Marguerite de Marillat'

'Merton Pride'

An old French cultivar grown for Louis XIV at Versailles

'Concorde'

'Catillac'

'Gormande'

'Packham's Triumph'

Quince

These large, golden, pear-shaped fruits have enjoyed a resurgence of interest in recent years. With their vanilla-like perfume, their growing presence on the show bench is most welcome.

JUDGES SCORE HIGHLY

Large, shapely fruits with unblemished skins and their eyes and stalks intact.

DEFECTS TO AVOID

Small, misshapen fruits and any that have damaged eyes, are blemished, lack stalks, or show evidence of rot around the stalk.

POINTS AVAILABLE

Condition	4
Uniformity	4
Size	4
TOTAL	12

PREP Cut the fruits from the tree, leaving the stalk intact. Judges will check for any rot forming around the stalk, so examine your fruits carefully before presenting.

PRESENT Cultivars that mature to a pear shape are best arranged in a circle on a plate with stalks poitning towards the centre. Stage those with a more apple-like shape as for apples (see pp.36–37).

JUDGING NOTES Exhibitors are normally required to show 6 fruits, or a minimum of 3 fruits at smaller shows. Unlike most other fruit classes, quinces can be shown either ripe or unripe. However, where the schedule does not specify ripeness, judges will always prefer cultivars that are ripe and in season over larger or more showy cultivars that have been picked too early.

Fully mature fruits should be yellow in colour

Furring of the skin is natural and need not be removed

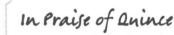

In Praise of Quince

Grown in Britain since the 13th century, quinces deserve to be more popular, not least because the spicy almond-vanilla-cream taste of cooked quince is a rare treat. The trees are self-pollinating so you need plant only one, and they will crop prolifically from an early age. Leaf blight can be a problem but the cultivar 'Serbian Gold' shows good resistance to it.

The calyx ends should be rounded and not split

The attractive leaves of the medlar can enhance a display

Ripe fruits are soft and have a slightly rancid smell

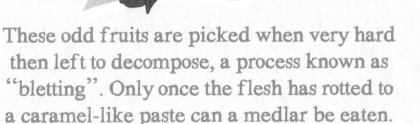

Medlars

These odd fruits are picked when very hard then left to decompose, a process known as "bletting". Only once the flesh has rotted to a caramel-like paste can a medlar be eaten.

PREP Medlars are ready to be picked when the stalk parts easily from the tree, but use secateurs or sharp scissors to cut fruit from the tree to leave the stalk intact. Handle the fruits as little as possible, by their stalks. If you intend to show bletted fruits, harvest and store them 2–3 weeks in advance of the show.

PRESENT Arrange the fruits in a circle in the centre of the plate, with the stalks facing inwards.

JUDGING NOTES Competitors are normally asked to show 10 medlars, or a minimum of 5 at smaller shows. Medlars may be shown either ripe or unripe, but ripe fruits are preferred.

JUDGES SCORE HIGHLY
Preferably ripe fruits with clean skins, stalks intact, and of a size characteristic of the cultivar.

DEFECTS TO AVOID
Fruits that are small for the cultivar and too firm or overripe, splitting, rotting, or with blotched skins.

POINTS AVAILABLE

Condition	3
Uniformity	3
Size	2
TOTAL	8

Grapes

Grape entries are more abundant in the glasshouse class, which is dominated at top shows by produce from aristocratic gardens. Entries to the outdoor class, however, are starting to grow thanks to hardier cultivars and a warming climate.

JUDGES SCORE HIGHLY

Large, complete, well-balanced bunches, of uniform size and shape, with fully ripe, uniform berries, of good colour and with a dense, intact bloom. Symmetry is preferred in glasshouse bunches.

DEFECTS TO AVOID

Small, unbalanced bunches, lacking uniformity in size or shape, with un- or overripe berries that are small, split, rotting, or diseased, and with little or imperfect bloom. Bunches should not be loose or so crowded as to have caused the berries to develop unevenly.

POINTS AVAILABLE

GLASSHOUSE

Condition	5
Size, shape, and density of bunch	5
Size and uniformity of berry	5
Colour	5
TOTAL	20

OUTDOOR

Condition	4
Size and uniformity of berry	4
Size, shape, and density of bunch	4
Colour	4
TOTAL	16

PREP Thin out bunches so that each individual grape has room to develop. Choose large, well-balanced bunches with full sets of grapes that have all ripened to a good size. Cut bunches with secateurs to leave approximately 50mm (2in) of stalk. In addition, glasshouse grapes should have a lateral shoot on either side of the stalk to form a T-handle. Handle carefully by the stalk to avoid damaging the bloom.

PRESENT Unless a method of presentation is specified in the schedule, glasshouse grapes are typically staged hanging from hooks screwed into a display board. Both winemaking and dessert cultivars of outdoor grapes may be shown on plates.

JUDGING NOTES Schedules normally require 1 bunch in a glasshouse class; entries of 2 bunches are permitted but need not score more points. For outdoor grapes, competitors are normally asked to show 2 bunches, or 1 bunch at smaller shows. Large, poor-quality bunches don't score as highly as smaller, good-quality ones. Glasshouse and outdoor grapes are judged separately, so make sure your entries are shown in the correct category, otherwise they risk disqualification.

Modern Vintages

New cultivars have made it possible to grow both wine and dessert grapes outdoors all over Britain. 'Lakemont' (pictured) is a muscat-like dessert cultivar that can be grown outdoors. While the red-wine grape 'Regent', bred specifically for cooler climates, will also grow well outdoors against a sunny wall.

Hang glasshouse grapes from hooks screwed into a board

Stems should be fresh and not withered

Glasshouse grapes can be shown in single or double bunches

Berries should show a good bloom with no spots or blemishes

Glasshouse Grapes

Stage outdoor grapes on plates

Try to match the shape of bunches as closely as possible

Outdoor Grapes

TOP TIPS

Birds and squirrels love grapes, so protect them with netting once the fruits form. Thinning out smaller fruits in the developing bunches with sharp, long-nosed scissors will ensure the grapes are perfectly formed for the show bench.

Cape Gooseberries

For sheer prettiness, this South American native is a welcome presence on the show bench. While they look like vivid orange-yellow sweets wrapped in paper cases, their tartness can be surprising.

JUDGES SCORE HIGHLY
Clean, dry, ochre-coloured, blemish-free husks (calyces), containing large, unblemished, ripe fruits of good colour.

DEFECTS TO AVOID
Immature, damaged, blemished husks (calyces), containing small, un- or overripe fruits of poor colour or condition.

POINTS AVAILABLE

Condition	4
Uniformity	2
Size	2
Colour	2
TOTAL	10

PREP Cut cape gooseberries from the plant when the calyces (husks) have turned an ochre colour and the fruits are bright orange-yellow. Unripe fruit may ripen if placed in a sunny, dry spot indoors. Take care not to squash or tear the husks.

PRESENT Arrange the fruits in a circle around the perimeter of a plate, or in symmetrical rows, with their husks intact and the stalk ends facing out. Do not reveal any of the fruits.

JUDGING NOTES Competitors are normally asked to show 20 fruits, or a minimum of 10 at smaller shows. Judges will open the husk of at least one fruit to check it is in good condition, ripe, and of a reasonable size.

What's in a Name?
Cape gooseberries are related to tomatoes and tomatillos, and have nothing at all to do with gooseberries. The Cape part of their name derives from the Cape of Good Hope, the region on the southern tip of Africa where they were cultivated by early European settlers.

Present fruits with the husks intact

Fruits should be a bright orange-yellow

Inedible papery husks cover the fruit

Rhubarb

Rhubarb is botanically a vegetable, but more commonly considered a fruit, and may appear in either category on a schedule. You may also find two classes listed: natural and "forced".

PREP Produce "forced" rhubarb by excluding light from the crowns, which "forces" plants to grow pink stalks a few weeks early (see Top Tip). Pick stalks of both types when they reach about 30cm (12in): hold them firmly at the base and twist off by hand. Trim the top foliage of natural rhubarb, leaving 75mm (3in) from the start of the leaf stalks. Do not remove the foliage of forced rhubarb. Trim off any bud scales at the base.

PRESENT Stage directly on to the bench, stalks arranged side-by-side with bases to the front. Do not tie the stalks together with raffia as it impedes the judging.

JUDGING NOTES Entrants are normally asked to show 3 stalks, or at least 2 at smaller shows. Forced and natural rhubarb are judged separately. Judges may break one stalk of each exhibit to check freshness and colour.

JUDGES SCORE HIGHLY
Fresh, firm, straight, long, tender stalks of uniform overall length and weight. Leaves of forced rhubarb should be small and undeveloped.

DEFECTS TO AVOID
Small, limp, crooked, stunted, tough, or damaged stalks. Any developed leaves in forced rhubarb.

POINTS AVAILABLE

NATURAL

Condition	3
Uniformity	3
Size	3
Colour	3
TOTAL	12

FORCED

Condition	4
Uniformity	3
Size	3
Shape	2
Colour	3
TOTAL	15

Foliage should be trimmed to about 75mm (3in)

stalks should be straight and roughly the same length and weight

stalk snapped in two to check for freshness and colour

Base should be free of bud scales

TOP TIP
Forced rhubarb is grown in the absence of light, and the tender, pink stems are sweeter than the greener natural version. To force rhubarb, cover the crown with a little straw and an upturned bucket or large pot, just before the shoots appear.

GROW TO SHOW
VEGETABLES

From leafy greens to robust roots, there's a vegetable to suit every would-be exhibitor, while adventurous souls may want to try something a little more giant...

Tips & Tricks
VEGETABLE CLASSES

Speak to any expert vegetable exhibitor and they will have their own secrets for success, whether it's a perfect compost recipe or a complicated watering schedule. Aside from the business of growing, however, you must also get to grips with the basics of exhibiting: when to harvest, how to prepare specimens, and how best to present.

Individual Vegetable Classes

PREPARATION Most specimens should be harvested as close to show day as possible. Alliums are the exception to this rule: onions, shallots, and garlic should be dug up a few weeks before show day, to allow time for the skins to dry (see pp.84–85 for more information). Trim or remove stalks, foliage, and side roots as required. If the specimens need to be washed, do so carefully, with a soft cloth and plenty of water. Do not use a hard brush, as this will damage the skin and spoil the specimen's appearance. Retain the natural "bloom" of the vegetable wherever possible. Under no circumstances should oils or similar substances be used in an attempt to enhance the appearance of the exhibit.

TRANSIT Pack entries carefully into cardboard boxes (or other suitable containers) lined with

Back to Black

After all the effort invested in growing and preparing your exhibits, don't neglect the staging. A neat arrangement can boost your chances of success in close competition, as it shows the judges you care. Try staging large root vegetables, such as carrots or leeks, on black cloth rather than laying them directly on the bench: a dark background can make colours look brighter and bolder. Some of the most experienced exhibitors take this method further, mounting their specimens on homemade angled boards to guarantee an eye-catching display.

paper or kitchen roll. Do not overfill the boxes, as the weight of the upper specimens may crush or bruise the ones below. Dividing the produce between several boxes will avoid the need to lift one overly heavy box around the show grounds. Giant vegetables pose their own transport challenges: depending on the entry's size, you may need to obtain the use of a van to transport your exhibit, and recruit a team of strong friends to help lift it into the van.

PRESENTATION Stage specimens as attractively as possible, on a plate, a dark-coloured board, or directly on the show bench, depending on size and type. Certain leafy vegetables, such as kale and chard, should be arranged in vases filled with fresh water. If necessary, leave the final trimming of stalks and leaves until just before judging, especially for vegetables with heads or stalk ends prone to discoloration.

Mixed Collections and Trug Exhibits

PREPARATION Check the schedule to confirm the quantities and requirements of any mixed collection and/or trug classes offered. Mixed collection classes tend to require a set number of specimens, while trug classes usually specify a minimum number of kinds or varieties of vegetables to be displayed.

TRANSIT Pack vegetables as for single cultivar exhibits (see left). Do not transport specimens to the show already in the trug, or attempt to lift a filled display trug by the handle, as this may damage both the trug and the veg within.

PRESENTATION Produce in mixed collections are usually presented as they would be for individual classes. Some exhibitors, however, like to stage their collections so that they are presented upright or raised at an angle, especially in a class featuring vegetables with long stems or roots. Trugs and baskets must be artfully arranged, as for fruit trugs (see p.15).

Attention to Detail

Vegetables exhibited in a trug collection should be prepared to the same standard as those featured in individual classes. Everything on display should be at the correct stage of ripeness or, in the case of alliums, be properly dried. It is a certainly a challenge to bring so many different types of vegetables to the peak of perfection all at the same time, which is why some growers consider the trug class to be the ultimate test of skill.

Tomatoes

Tomatoes were once thought poisonous and given the name "wolf peach", but are now the most common home-grown fruiting vegetable in Britain. From tiny cherries to mighty beefsteaks, there are multiple tomato classes to choose from.

JUDGES SCORE HIGHLY
Shapely, ripe but firm fruits of the correct size for the category. They should be well-coloured and blemish-free, and should have fresh calyces attached.

DEFECTS TO AVOID
Fruits that are too large or small for their category, misshapen, unripe or overripe, blemished, or of a dull colour. Entrants should also not enter green-backed specimens, those with evidence of pest or disease damage, or lacking calyces.

POINTS AVAILABLE

LARGE AND TRUSS
Condition	5
Uniformity	3
Size	3
Shape	2
Colour	2
TOTAL	15

MEDIUM AND SMALL
Condition	5
Uniformity	4
Size	3
Shape	3
Colour	3
TOTAL	18

PREP Select uniform specimens of the right shape, size, and colour for the cultivar. When harvesting, leave the calyx (green husk) and a short portion of stalk attached. Do not select any overripe fruits, or those with a hard "green back" around the calyx.

PRESENT Stage on a plate with the calyces uppermost, if possible. Smaller cultivars may be unable to sit upright and small rings or sand can be used to steady their bases. Trusses may be staged with or without a plate.

JUDGING NOTES If multiple tomato classes are offered, these typically specify an ideal size, which you should aim for when selecting fruits to enter. Always choose your category based on the standard size of the cultivar; entering undersized specimens of a typically large cultivar into a smaller category will not succeed. The number of specimens that must be submitted tends to vary between shows and between classes; generally, smaller classes will require more specimens than larger ones. Check the schedule before entering. Small-fruited tomatoes, such as cherry or currant tomatoes, should not exceed 35mm (1³/₈in) in diameter. Medium tomatoes, such as plum or salad tomatoes, should be approximately 60mm (2¹/₂in) in diameter. Large tomatoes, such as beefsteak, should measure about 75mm (3in) in diameter. Truss tomatoes are exhibited on the vine. They should ideally have at least one third of the fruit fully ripe.

Staging Small Tomatoes
Clean any dirt with a soft cloth. If you want the tomatoes to sit upright, place them on a bed of sand or on metal staging rings.

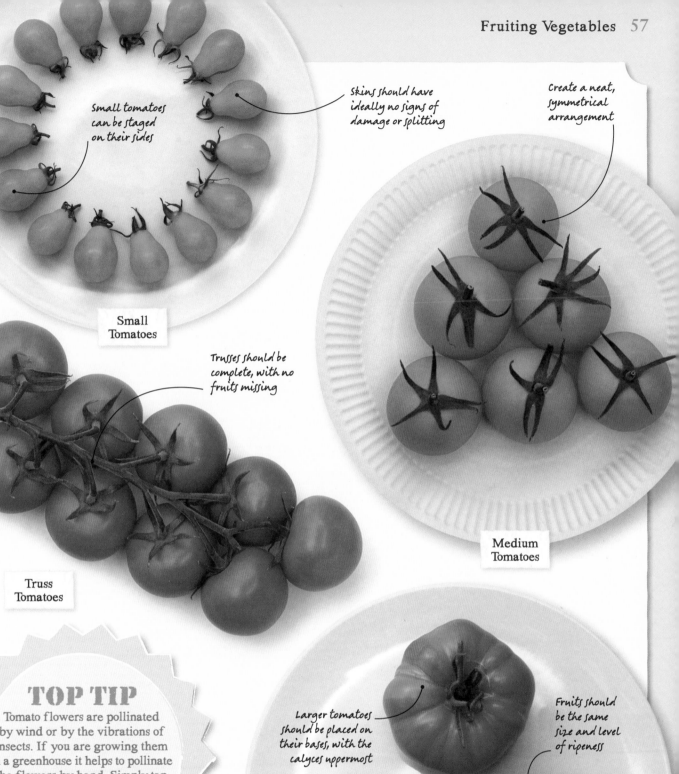

Small tomatoes can be staged on their sides

Skins should have ideally no signs of damage or splitting

Create a neat, symmetrical arrangement

Small Tomatoes

Trusses should be complete, with no fruits missing

Medium Tomatoes

Truss Tomatoes

Larger tomatoes should be placed on their bases, with the calyces uppermost

Fruits should be the same size and level of ripeness

Large Tomatoes

TOP TIP

Tomato flowers are pollinated by wind or by the vibrations of insects. If you are growing them in a greenhouse it helps to pollinate the flowers by hand. Simply tap them when the pollen is dry and repeat for two or three days on the same flower clusters.

Sweet Peppers

Botanically, sweet and chilli peppers are both members of the capsicum family. However, due to their larger size and mild flavour, sweet or bell peppers usually appear in their own show class.

JUDGES SCORE HIGHLY
Fresh, brightly coloured fruits, uniform in the colour appropriate to the cultivar, free from blemishes, of good size and shape. Fruits should also show clear evidence of a fresh stalk.

DEFECTS TO AVOID
Immature, partially coloured fruit. Misshapen fruit displaying poor pollination or having no stalk.

POINTS AVAILABLE

Condition	5
Uniformity	3
Size	2
Shape	2
Colour	3
TOTAL	15

PREP Select uniform fruits that are of the right shape and colour for the cultivar. Leave a portion of stalk on each fruit.

PRESENT Arrange on a plate. Depending on the shape and size of the cultivar, peppers may be laid flat in a circle, with stalks directed outwards, or stood with stalks pointing up.

JUDGING NOTES Entrants are normally asked to supply 3–6 sweet peppers, or a minimum of 2 at smaller shows. You may show either immature, green fruits (as long as they are fully formed) or mature fruits that have reached the coloured stage. However, mature specimens are preferred.

TOP TIP
Like tomatoes, peppers need high light levels and warm conditions. As a result, the best crops are produced in a greenhouse, or grown at the base of a south-facing wall in borders or grow-bags.

Coloured peppers are preferred to immature green specimens

Sweet peppers can be presented either standing or laid on a plate

Choose specimens of a uniform colour and size

Chilli Peppers

When Columbus first came across chillies he thought they were a fiery type of peppercorn, and the name stuck. Despite the huge variety of chillies, all are judged by the same criteria.

PREP Select uniform fruits that are of the right shape and colour for the cultivar. When harvesting, leave a portion of stalk on each fruit.

PRESENT Depending on the length and thickness of the cultivar, chilli peppers may be arranged on a plate in a circle, or laid side-by-side on the table.

JUDGING NOTES Competitors will normally be asked to supply 6 chilli peppers, or a minimum of 3 at smaller shows. You may show either immature, green fruits (as long as they are fully formed) or mature fruits that have reached the coloured stage. However, mature specimens are preferred.

JUDGES SCORE HIGHLY
Fresh, well-developed fruits with a lustrous, bright, uniform colour according to cultivar. Skins should be free of blemishes. Stalks attached.

DEFECTS TO AVOID
Fruits that are limp, soft to the touch, or that show variation in maturity and colour.

POINTS AVAILABLE

Condition	4
Uniformity	3
Size	3
Shape	2
Colour	3
TOTAL	15

Long chillies, such as 'Joe's Long', can be presented directly on to the show bench

Points are awarded for full-size specimens

Scotch Bonnet Chillies

Cayenne Chillies

stalks should remain attached

Pinching and Picking

Once the stems reach a height of around 30cm (12in), pinch out the tips to encourage bushy growth so that there are more stems producing flowers and fruit. At the same time, stake the plants to provide support. Pick fruits regularly once they are glossy and smooth, which encourages further fruit production.

The Urban Gardener

My name's Sarah Plescia, I'm from south London and I've been competing in my local urban village show for almost a decade.

BEAUTY IN THE EYE OF THE GROWER

I won first prize in the marrow class this year for my lovely big marrow. I've also grown a fantastic lettuce, some peppers, some really nice beans, and some very pretty shallots. Not all my vegetables were best in class, but I don't mind. They're not all perfect, but they are all beautiful in my opinion.

That's the best part of coming to the show: to look at all the beautifully grown things. There are always some spectacular exhibits and the entries just seem to get better each year. I heard someone this year had travelled with a whole host of beautiful veg from the other side of London – on public transport, no less! And the effort doesn't stop there. Watching people set up their displays, it looks like flower arranging for vegetables. Take shallots, for example. In real life, no one bothers making their shallots look presentable. But the night before the show, I stayed up till 1.30 in the morning dressing my shallots – carefully tying string round these miniature onions. It's ridiculous when you think about it. But it's also wonderful. Once you're done, your display looks like a little shallot Zen garden and the effect can be phenomenal. You take pride in the appearance of your exhibit.

GETTING STUCK IN

I've been competing in the show for nine years or so; before that I would often come down to see the exhibits. Time and time

> "People love village shows because they are charmed by the oddness of the event."

Shallot... Then Not!
The unfortunate theft of one of Sarah's bulbs spoilt the Zen perfection of her shallot display this year, but it hasn't made her bitter: she likes to think someone was simply one shallot short of a good stew.

> ## "I don't mind if I end up with vegetables that look a bit ropey and aren't show-ready. I'll just eat them instead!"

Good for the Plate
Prize-winning or not, all Sarah's entries end up eaten.

again, I used to look at the rhubarb on the show table and think "well, that does look nice, but my rhubarb crop looks better than that". I kept saying this year after year, never doing anything about it. Finally, one year I put my name down on the list and ended up winning the rhubarb class! After that, I couldn't stop myself.

FOR THE LOVE OF LETTUCE
Lettuce is probably my favourite vegetable to grow, because they are so ordinary in so many ways. You can easily buy a head of lettuce in a supermarket. But it's tricky to produce an absolutely immaculate, beautiful-looking lettuce. And yet, when you manage it – when the slugs haven't got to it, and you've kept

the bugs from dropping in it, and your allotment neighbour hasn't accidentally strimmed it – what you're left with is almost like the head of a flower, it's so perfect and beautiful. Not only that, but it's such a fragile, vulnerable crop. Maybe that's why they're so great: you get that immense achievement of raising something so fundamentally vulnerable.

City Institution
Held in a park in a built-up area of south London, Sarah's local show has become a summertime institution and now attracts more than 100 competitors from many different parts of the city.

Aubergines

Rising temperatures mean aubergines are becoming increasingly visible on the show bench. Deep purple fruits are popular, but lilac-striped or creamy white cultivars are also available.

JUDGES SCORE HIGHLY

Large, shapely, bright, well-coloured, solid fruits, free from blemishes and with fresh calyces and the natural bloom of the skin intact.

DEFECTS TO AVOID

Fruits that are small, misshapen, shrivelled, dull or poorly coloured.

POINTS AVAILABLE

Condition	5
Uniformity	4
Size	3
Shape	3
Colour	3
TOTAL	18

PREP Harvest carefully, retaining a good portion of stalk, and take care to preserve the natural skin condition; if dirty, wash them with a soft cloth and plenty of water.

PRESENT Lay large fruit side-by-side on a plate, with stalks directed upwards. Present smaller fruits in a round, with stalks pointing inwards. Fresh leaves can add interest to an exhibit.

JUDGING NOTES Competitors normally present 3 specimens, or a minimum of 2 at smaller shows. Categories for smaller specimens may request a higher quantity.

TOP TIPS

Aubergines need constant temperatures of 25°C (75°F), so for best success grow in a greenhouse or against a sunny wall. Allow only 3–4 fruits per plant, so they reach an optimal size. Feed with high-potash fertiliser once the fruits form.

Present specimens with stalks and calyces (green husks) attached

Judges look for a high natural shine

Choose large, shapely fruits of a comparable size

Sweet Cravings

Corn is a hungry crop, so apply a general fertiliser before planting. They need an open, sunny site and deep, well-drained soil. Take care when planting out seedlings, as they resent root disturbance.

Leave a portion of stalk intact

Peel back the husk to reveal one quarter of the grains

Present cobs filled with full rows of tender grains

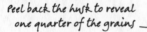

Sweetcorn

Sweetcorn is an amazingly versatile crop, used to produce everything from laundry starch to explosives! Judges reward cobs that are full to the brim with ripe, tender grains.

PREP Sweetcorn is ready to harvest when the tassels at the top of the plant turn brown. Choose well-filled cobs of a uniform size.

PRESENT Trim the stalks just before presenting. Cobs should be displayed with around one quarter of the grain exposed. To achieve this presentation, grip the top of one husk and pull down sharply, from the tip to the base.

JUDGING NOTES Competitors normally need to present 3 cobs, or a minimum of 2 at smaller shows. Judges will pull down the husks further to reveal the grains fully.

JUDGES SCORE HIGHLY
Long, fresh, cylindrical cobs, well set throughout, with straight rows of undamaged, plump, tender grains.

DEFECTS TO AVOID
Cobs that are not fresh, are unduly tapered, have irregular rows of grain or are badly set. Husks that are shrivelled and straw-coloured.

POINTS AVAILABLE

Condition	5
Uniformity	4
Size	3
Set of grain	3
Colour	3
TOTAL	18

Broad Beans

Sowing broad beans in late autumn is an act of faith in the coming year. With the right care and conditions, you could have early shoots before Christmas, and be harvesting pods by May.

JUDGES SCORE HIGHLY
Fresh, well-filled pods with stalks and clear unblemished skins; tender seeds of good size.

DEFECTS TO AVOID
Pods that are not fresh or are blemished, imperfectly filled, or contain seeds that are not tender.

POINTS AVAILABLE

Condition	5
Uniformity	3
Size	3
Shape	2
Colour	2
TOTAL	15

PREP Choose long, blemish-free specimens of a uniform length. Use scissors to cut beans from the vine, leaving a good portion of stalk on each pod. Check spare pods to assess interior freshness.

PRESENT Stage the pods either on a plate or directly on the bench, in a row with tail ends facing the front of the bench.

JUDGING NOTES Entrants are normally asked to submit 9 pods, or a minimum of 5 at smaller shows. Judges will break one pod to determine freshness and check that seeds are not split, detached from the pod, do not have dark "hilums" (eyes), and are free of signs of pest and disease damage.

Bean Feast for the Crow?
There's an old saying about sowing broad beans: "one for the rook, one for the crow, one to rot, and one to grow". This comes from centuries of rueful experience, when growers would plant four seeds for every one likely to survive, expecting the others to be eaten by pests or rot in cold, wet conditions. Even today, it's still a good idea to improve the odds by sowing a few extra seeds at the end of rows, in order to have some spare plants for filling gaps where seeds haven't germinated.

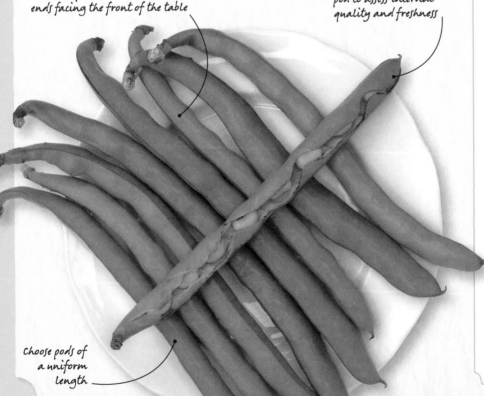

Present pods in a row, with tail ends facing the front of the table

Judges will snap one pod to assess internal quality and freshness

choose pods of a uniform length

Fresh Beans

Judges will snap open one pod to assess internal condition

Shelling Beans

Seeds of shelling beans should be fresh, neither immature nor overly mature

French Beans

Also classed as dwarf, climbing, or shelling, French beans aren't even French: originally from the Americas, they were brought to Europe by the Spanish in the 1500s.

PREP Use scissors to cut specimens from the vine and check inside some spare pods to assess the condition of the seeds.

PRESENT Stage as you would broad beans (see opposite). Longer beans may also benefit from being displayed on a dark board.

JUDGING NOTES Entrants normally submit 9 pods, or around 5 at smaller shows. Judges will break one pod from each exhibit to determine freshness and condition. Climbing and dwarf French beans can be difficult to distinguish on the plate and for competition purposes they are considered the same. However, if you are growing a shelling variety of French bean (i.e. one that is not eaten whole but grown for the edible seeds inside), you may find a separate class for "Beans, shelling, other than broad". Enter as a French bean if there is no such class in the schedule.

JUDGES SCORE HIGHLY
Straight, fresh, tender pods of good colour, size, and even length, with no outward sign of seeds, and having stalks and uniform tails.

DEFECTS TO AVOID
Pods that are dull, pale, misshapen, shrivelled, tough, stringy or with prominent seeds. In shelling types, dry or imperfectly filled pods of poor colour, or having dry seeds.

POINTS AVAILABLE

Condition	5
Uniformity	3
Size	3
Shape	2
Colour	2
TOTAL	15

Runner Beans

The ideal length of bean for normal competition is 25cm (10in). If your runners are longer, why not enter a longest bean competition? The UK record currently stands at 91.4cm (36in).

JUDGES SCORE HIGHLY
Long, flat, straight, fresh young pods, uniform in colour, shape, and length.

DEFECTS TO AVOID
Stringiness, "bottlenecked" or otherwise misshapen pods, and large seeds clearly showing.

POINTS AVAILABLE

Condition	5
Uniformity	4
Size	3
Shape	4
Colour	2
TOTAL	18

PREP Use scissors to cut beans from the vine, leaving a good portion of stalk on each pod. Check some spare pods to assess condition and interior freshness.

PRESENT Neatly lay the pods side-by-side on a plate or dark board, stalks at one end and tails at the other, with tails facing out.

JUDGING NOTES Competitors are normally asked to submit 9 beans, or a minimum of 5 at smaller shows; if submitting as part of a vegetable collection you may need to show 12 beans. Judges will break one pod to determine freshness; they will be looking for a clean, crisp snap and no evidence of stringiness.

TOP TIP
Runner beans are thirsty plants. For a moisture-retentive planting space, dig a trench 45 x 60cm (18 x 24in), line the bottom with newspaper at least 3 sheets thick, and fill with compost in 10cm (4in) layers, firming as you go to prevent air pockets.

Trim your bean stalks to a uniform length

Pods should be flat, with minimal outward sign of seeds

Choose pods as long and straight as possible

Peas

Our love of peas goes back a long way and it was British breeders who first developed strains of "garden" peas to be eaten fresh. In show, full pods with a fresh bloom win prizes.

PREP Gather pods using scissors, cutting them with approximately 25mm of stalk attached. Hold the peas by the stalk, so you don't rub off the natural waxy bloom, as the judges will be awarding marks for that. Check for internal damage and the number of peas by holding pods up to a strong light.

PRESENT Peas are always presented for competition in the pods, never shelled. Arrange on a plate or on a dark display board, side-by-side facing the same way, with the tail ends at the front.

JUDGING NOTES Exhibits will require 9 specimens, or at least 5 at smaller shows. Sugar snap and mangetout peas are judged by the same criteria, except mangetout should be flat with undeveloped seeds and sugar snaps should have a fleshy texture and snap cleanly and easily.

JUDGES SCORE HIGHLY
Large, long pods with the exterior bloom intact and a good set of plump, tender peas.

DEFECTS TO AVOID
Small, poorly coloured pods, overdeveloped peas, and any signs of disease. No maggots!

POINTS AVAILABLE

Condition	6
Uniformity	4
Size	4
Shape	4
Colour	2
TOTAL	20

Handle with Care

Medwyn Williams has won 10 gold medals at Chelsea for his veg. His advice for protecting peas in transit is to pack them in fresh nettle leaves, which preserve and even enhance the bloom. Medwyn's recommended cultivar is 'Show Perfection', which consistently bears 10 peas per pod.

Pods should be large and long with a smooth exterior

Judges will open one pod to observe the contents

free
courgettes
from our
allotment

Cucumbers

You may find four different classes for cucumbers in a show schedule, depending on size, method of cultivation, and whether they are gherkin species – and if you count recent arrivals cucamelons, that makes a fifth category too.

JUDGES SCORE HIGHLY
Fresh, young, green, tender, blemish-free, straight specimens. They should have short handles and be of a uniform thickness.

DEFECTS TO AVOID
Old, yellowing, crooked, or soft fruits with long handles that are of irregular thickness, showing pest or disease damage, and any that are marked by contact with the stalks or leaves.

POINTS AVAILABLE

GREENHOUSE

Condition	5
Uniformity	4
Size	3
Shape	3
Colour	3
TOTAL	18

OUTDOOR, SMALL, AND GHERKIN

Condition	5
Uniformity	3
Size	2
Shape	2
Colour	3
TOTAL	15

PREP Cucumbers are ready for picking when they are a fresh green colour, with a completely developed flower end, a well-shaped barrel (main body), and a short handle (thin end). Cut fruits from the vine with a sharp knife, leaving a short section of stalk. Handle carefully to avoid damaging the bloom on the rind of the barrel. Flowers need not remain attached to specimens.

PRESENT Larger cucumbers may be staged directly on to the bench or on plates, while smaller varieties are best shown on plates. All cucumbers should be lined up and facing the same direction.

JUDGING NOTES Entrants are normally asked to show 2 outdoor or greenhouse specimens. For mini, small, and pickling cucumbers and gherkins, 6 specimens are usually specified or a minimum of 3 at smaller shows. Greenhouse cucumbers should be 250mm (10in) or more in length; outdoor cucumbers can be smaller. Small, pickling, and gherkin cucumbers should be 100–200mm (4–8in) in length. Cucamelons should be entered into the class for miscellaneous vegetables. Entries will garner up to 12 points and judges are looking for ripe, firm, blemish-free, well-coloured fruits of uniform shape and size.

Yellow Monsters

Cucumbers feature in both the heaviest and longest categories of giant veg competitions. They are too sour and fibrous to be edible, but that isn't really the point. The records currently stand at 13kg (28½lb) for weight and 107cm (42in) for length.

What on Earth is a Cucamelon?

These grape-sized oddities have skins patterned like watermelons and taste like cucumber with a dash of lime. Also known as sour gherkins, they are the fruits of a species of vine from South America that is a close relation to cucumbers.

Outdoor cucumbers may be much smaller than indoor types

Outdoor types are also known as ridge cucumbers due to their ridged skins

Cucumbers grown indoors should be at least 250mm (10in) long

Cucumbers should be straight and blemish-free

Outdoor Cucumbers

Show cucumbers with or without their flowers

TOP TIP

The Victorians used special glass cylinders to straighten their cucumbers. Today, it's more usual to straighten specimens by hand: when they are 15cm (6in) long, gently bend the barrels just before watering and after the sun has shone on them.

Greenhouse Cucumbers

Courgettes

While technically a summer squash, courgettes are usually shown in their own class. Don't fret if your courgettes outgrow the strict size criteria. Why not enter them into a marrow class instead?

JUDGES SCORE HIGHLY
Young, tender fruits, uniform in shape and colour, and which match the ideal size criteria for their type (see Prep). Specimens can be of any colour but must be well-matched.

DEFECTS TO AVOID
Old, misshapen, or blemished specimens, ill-matched or damaged by pests or disease. Wilted flowers.

POINTS AVAILABLE

Condition	4
Uniformity	3
Size	2
Shape	3
TOTAL	12

PREP Time your growing season carefully, as the ideal size for courgettes shown in competition is quite specific: standard varieties should be 150mm (6in) long and 25–35mm (1–1$\frac{1}{3}$in) in diameter, while round cultivars should be 75mm (3in) in diameter. You might want to consider taking a ruler out to the courgette patch.

PRESENT Exhibit courgettes flat on a plate, stalk end facing the front and with fresh flowers attached, if possible; judges look out for flowers as they can be an indicator of freshness.

JUDGING NOTES Competitors are normally asked to show 3 fruits, or at least 2 at smaller shows.

TOP TIP
Don't panic if early-season courgette flowers don't seem to be setting any fruit: these are the slender-stemmed male flowers needed to pollinate the females. Female flowers have embryo fruits and appear once the weather improves.

Specimens with flowers attached show freshness

The skin must be blemish-free

Courgettes can be any colour as long as they are well matched

Present directly on the show bench, not on a plate

Make sure the marrows are clean

Monster Marrows

If you want to try growing a record-breaking giant marrow, you will need a good deal of upper-body strength or friends willing to help lift it – the current UK record for heaviest marrow is 77.56kg (171lb).

Marrows

While giant marrow specimens are the stuff of village show legend, your crop needn't be so extreme. Most shows offer classes for regular-sized specimens – no forklift trucks required.

PREP Choose young, uniform specimens with a tender skin. Typical varieties are cylindrical with blunt ends, and should be no more than 350mm (14in) in length. Round cultivars should have a circumference of 500mm (19½in). Exclude any specimens that look too old: after all, if you wouldn't want to cook them, they do not belong on the show bench.

PRESENT Wipe clean before presenting. Lay directly on the show bench, not on a plate.

JUDGING NOTES Regular marrow classes usually call for 2 specimens; "giant" competitions generally ask for only 1. Green-striped marrows are traditional, but other colours are allowed.

JUDGES SCORE HIGHLY
A well-matched pair of fresh, young, unblemished fruits.

DEFECTS TO AVOID
Mature, oversized, blemished, or misshapen marrows.

POINTS AVAILABLE

Condition	6
Uniformity	4
Size	3
Shape	2
TOTAL	15

Trim stalks neatly

Fruits should have a good shape for the cultivar

Choose young but fully formed fruits to display in competition

Summer Squash

Winter Squash

Skin should be hard, with no spots or marks other than the natural colouring

Squash

Pattypan, crookneck, vegetable spaghetti: the names of squashes can be as wonderfully weird as the shapes and sizes they come in, from frilly flying saucers to warty lightbulbs.

PREP Summer Squash Select young, tender fruits, normally not more than five days after flowering. Take care not to mark the flesh when cutting from the vine, and retain a portion of stalk. Winter Squash Select fully coloured, mature fruit of a size appropriate to the cultivar. Retain a portion of stalk.

PRESENT Summer Squash Specimens should be staged flat, either on a plate or board, or directly on to the table.

Winter Squash Stage directly on the table, or on a board or cloth, depending on size; winter squash are too large for plates.

JUDGING NOTES Summer Squash 3 specimens are usually specified and may be shown with or without flowers attached. Although courgettes and marrows are technically summer squash, for the purposes of competition they are placed in separate classes. Winter Squash Competitors need only enter 1 specimen.

JUDGES SCORE HIGHLY
Shapely, large, firm fruit of even colour and ripeness. Stalk attached.

DEFECTS TO AVOID
Soft or blemished skins, uneven ripening, or missing stalks.

POINTS AVAILABLE

SUMMER	
Condition	4
Uniformity	3
Size	2
Colour	3
TOTAL	12

WINTER	
Condition	4
Size	3
Colour	3
TOTAL	10

Pumpkins

Pumpkins are an undoubted highlight of late-season horticultural shows – especially giant specimens. Nothing wows a crowd like a weigh-in for a "heaviest pumpkin" contest.

PREP Choose a mature, well-formed specimen of good, deep colour. A mature, harvest-ready pumpkin should have a hard rind and stem, and sound hollow when tapped. Take care not mark or bruise the skin when testing for maturity. Leave a good length of stalk of up to 10cm (4in) when cutting from the vine, and use a sharp knife or pair of secateurs to ensure a clean cut.

PRESENT Pumpkins are usually placed directly onto the table, but you could present on a large dark board or black cloth to give the specimen maximum presence.

JUDGING NOTES Entrants only need to show one specimen. Check the definition of pumpkin given in your show schedule to ensure you're not entering a squash into the wrong class.

JUDGES SCORE HIGHLY
Well-formed, large, shapely fruit that are firm, of good colour and ripeness, with stalk attached.

DEFECTS TO AVOID
Misshapen, soft, unevenly ripened fruit with a blemished or marked skin, or lacking its stalk.

POINTS AVAILABLE

Condition	4
Size	3
Colour	3
TOTAL	10

Fruits should always have a substantial stalk attached

Rind colour can range from yellow-orange to almost maroon, according to cultivar

Giants of Legend
There are some tall stories told about giant pumpkins. One grower claimed the secret to success was to induce a giant "beer gut" by feeding the pumpkin six pints of beer a day. Another story tells of some children who broke into an allotment to steal a giant pumpkin for Hallowe'en, only to injure themselves in a failed attempt to lift it.

A Weird World of Pumpkins & Squash

Don't be blinkered by the ubiquity of the butternut – there's a wonderfully varied range of pumpkins and squash available to grow.

Some squash naturally develop warts and it has no effect on flavour

'Yellow Warted'

Sunburst is a summer squash grown and cooked like courgettes

'Pattypan Sunburst'

'Caspar White'

Pumpkins and squash are members of the Cucurbit family of vegetables, along with courgettes and marrows – all are very hungry and thirsty crops.

'Mars'

This winter squash gets its name from markings that resemble bat wings

'Batwing'

'Green Turk's Turban'

'Summer Ball'

'Triamble'

'Yellow Crookneck'

'Hooligan'

Squashing the Opposition

Scales are the ultimate arbiter at the Giant Pumpkin competition. Sheer girth is not always an indicator of weight and the weigh-in is the moment of truth when competitors find out if their pumpkin is dense enough to win "heaviest in show".

Heavy Goods
A grower of giant pumpkins can travel far and wide to compete and needs a sturdy means of transportation.

Heavy Lifting
Positioning pumpkins is back-breaking work and true giants need a forklift truck to move them.

The Giant Pumpkin Grower

I'm Matthew Oliver, I work as a horticulturist at RHS Hyde Hall in Essex and I currently hold the record for the heaviest pumpkin grown outdoors in Britain.

THE SEED OF AN IDEA

I got into horticulture through family. My parents and grandparents had their own allotments and from a young age I was exposed to allotment life.

I'm now in charge of the fruit and veg garden at RHS Hyde Hall. When I took over we were already growing a range of pumpkins for autumn displays. Each year you try to make the display bigger and better, and we found it was the giant ones that really grabbed people's interest. This year I made a conscious decision: we're growing a proper giant!

> "Growing giant veg, you're not really competing against each other – you're competing against the scales."

First I had to source some decent seed. You can't use any old seed to grow a giant pumpkin. Hobby growers have been developing the 'Atlantic Giant' strain for the past 30 or 40 years, cross-breeding and selecting the biggest ones year-in, year-out.

Then I got in touch with the online community of giant growers to find out how it's done. Social media has had a massive impact on the giant veg community, and for "grow to show" more generally, because suddenly gardeners can get connected.

GREEN-EYED GROWING

You have to start early in the year to grow a proper monster. Sow mid-April and plant out mid-May. By the end of June the plant will have put on enough leaf growth and you can pollinate a flower by hand to start the pumpkin growing.

Sun Block
To prevent it from ripening too early, the pumpkin was protected from the sun's heat by wrapping it in a white sheet and green shade netting.

MONSTER STATS

COST OF SEED	£1,250
DAYS TO GROW	149
PEAK GROWTH RATE	20.4kg (45lb) per day
FINAL CIRCUMFERENCE	442cm (14½ft)
FINAL WEIGHT	605kg (1,334lb)

Precious Seed
The seed set a record as the most expensive in the world, and was over three times the size of normal pumpkin seed.

You need to prep the ground well with lots of compost, and you need to take soil tests to check you've got a good balance of nutrients. People say gardening is about green fingers but I say it's about green eyes: it's about interpreting what you're seeing. I had a big problem with magnesium deficiency. I was getting lots of yellowing in the leaf and I cured that through spraying with Epsom salts once a week.

FEEDING THE MONSTER

You have to feed and water every day. And you need to give the same amount at the same time. If you start being inconsistent, miss a few days' watering, you get these fluctuations and the pumpkin might split and you can't take a pumpkin that's damaged to a weigh-in. That's why it's a huge drain on your time – you can't go off on holiday.

My girlfriend thought it was quite fun but I don't think she'd be keen for me to do it every year. Even top growers have a year out. Life gets in the way.

All Hail the Great Pumpkin
When Matt took over he wanted to supersize the pumpkin patch and bring a true giant to Hyde Hall.

"**The biggest danger is that the pumpkin grows too quickly and splits, literally explodes on the plant!**"

Asparagus

Plan ahead if you want to show asparagus as it takes three years from planting to a decent crop. Spears are harvested for six weeks from mid-spring, so you'll need to enter an early show.

JUDGES SCORE HIGHLY

Fresh, long, straight, plump spears with well-closed scales and a good colour appropriate to the cultivar.

DEFECTS TO AVOID

Stems that are not fresh, or are short, crooked, thin, shrivelled or dull-coloured, or with open scales.

POINTS AVAILABLE

Condition	5
Uniformity	3
Size	3
Shape	2
Colour	2
TOTAL	15

PREP Harvest spears as close to show day as possible, to ensure maximum freshness. Choose straight, uniform spears, and cut them with a sharp knife 2cm (³/₄in) below the soil surface. Each specimen should have roughly the same length of presentable spear; you can trim them more exactly at the show. Make sure they are clean, but take care not to remove the natural bloom.

PRESENT Stage spears in tight formation side-by-side on a plate, with the cut ends to the front.

JUDGING NOTES Competitors are normally asked for 6 spears of asparagus, or a minimum of 3 at smaller shows. Judges may snap one spear to test for freshness and inspect internal condition. Purple cultivars are judged according to the same criteria as green types.

Spear tips should have well-closed scales

Cut specimens to a uniform length

Ensure the cut ends are as fresh as possible

Waiting Game

Establishing an asparagus crop takes time. Plant "crowns" (rootstocks) in the spring of year one, but leave the spears to grow and cut back the stems in autumn. The next year you will be able to cut a few small spears, but it is not until the third year of growth that spears may be freely harvested.

Globe Artichokes

Globe artichokes are related to thistles and the "globe" is actually a large flower bud. Plants produce multiple buds, but it is the main "king" bud that judges will be looking for.

PREP As the plant begins to produce buds, remove any lateral (side) buds so that all energy goes into producing the central king bud. Keep watering levels consistent throughout the summer, as spells of drought will result in smaller buds. Harvest the king bud when the scales are close to opening but still compact. Carefully cut it from the plant with a sharp knife, and trim the stalk to about 5cm (2in).

PRESENT Stage the artichoke buds either directly on to the show bench with stalks to the front, or on a large plate with stalks pointing to the centre.

JUDGING NOTES Competitors are normally asked to submit 2 specimens, or 1 artichoke at smaller shows. Highest marks will be awarded to well-grown specimens with symmetrical heads and closely knit scales.

JUDGES SCORE HIGHLY
Large, heavy, shapely, closed buds composed of plump, solid, fleshy scales with minimal signs of damage to the exterior.

DEFECTS TO AVOID
Buds that are small, lightweight, irregular, or loose, or have thin or shrivelled scales. Buds that are notably different in size. Stalk that is too long or too short.

POINTS AVAILABLE

Condition	5
Uniformity	3
Size	2
Shape	2
Colour	3
TOTAL	15

Heads should be large and plump, with tight scales

Scales should be near to opening, but still closed

Trim any leaves from the stalk

Scales should be fleshy and fresh

Leeks

Leeks are amongst the biggest stars of the show bench and have their own distinct terminology to describe the parts that are judged (see box below). You will find three classes: blanched, intermediate, and pot leeks (the short fat ones).

JUDGES SCORE HIGHLY

Dark green, turgid flags; a tight button; a firm, solid barrel with a good blanch; a sound root plate.

DEFECTS TO AVOID

Discoloration of the blanch; a barrel that is not in size; a thin, tapering, or bulbous barrel; signs of disease.

POINTS AVAILABLE

Condition	6
Uniformity	4
Size	4
Shape	3
Colour	3
TOTAL	20

TOP TIPS

To ensure straight stems, keep transplanted leeks upright by supporting them with split canes attached to the stems using specialist leek clips available online. Blanch by wrapping them in cardboard or black damp-proofing material.

PREP Clean stems, leaves, and roots by flushing with water, but avoid washing any soil particles between the leaves. Tease out the roots but do not trim them. Avoid stripping too many outer leaves as this can expose ribbing on the stem, which judges will mark down.

PRESENT Stage leeks either directly on the bench or on a blackboard, with their roots to the front of the table.

JUDGING NOTES Blanched leeks must measure more than 350mm (13¾in) from base to button. Intermediate leeks must measure not less than 150mm (6in) and not more than 350mm (13¾in) from base to button. Pot leeks must measure no more than 150mm (6in) from base to button and the barrels should have a good volume to them. Otherwise, all types are judged by the same criteria and points scheme. You will usually be asked to display 3 specimens of blanched and intermediate leeks, and 2 specimens of pot leeks. In close competition, judges may measure the length and girth of each exhibit to aid selection.

Know Your Leek Terms

The terms used in a show schedule to describe the various parts of a leek can be confusing to the uninitiated. This quick guide will help you to decipher.

Flags – the name for the leaves of the plant, which are presented in their entirety

Button – the point on a leek where the lowest leaf (or flag) breaks the circumference of the blanched stem (or barrel)

Barrel – the name for the main shaft of a leek stem

Beard – the name for the roots of a leek, which must not be trimmed

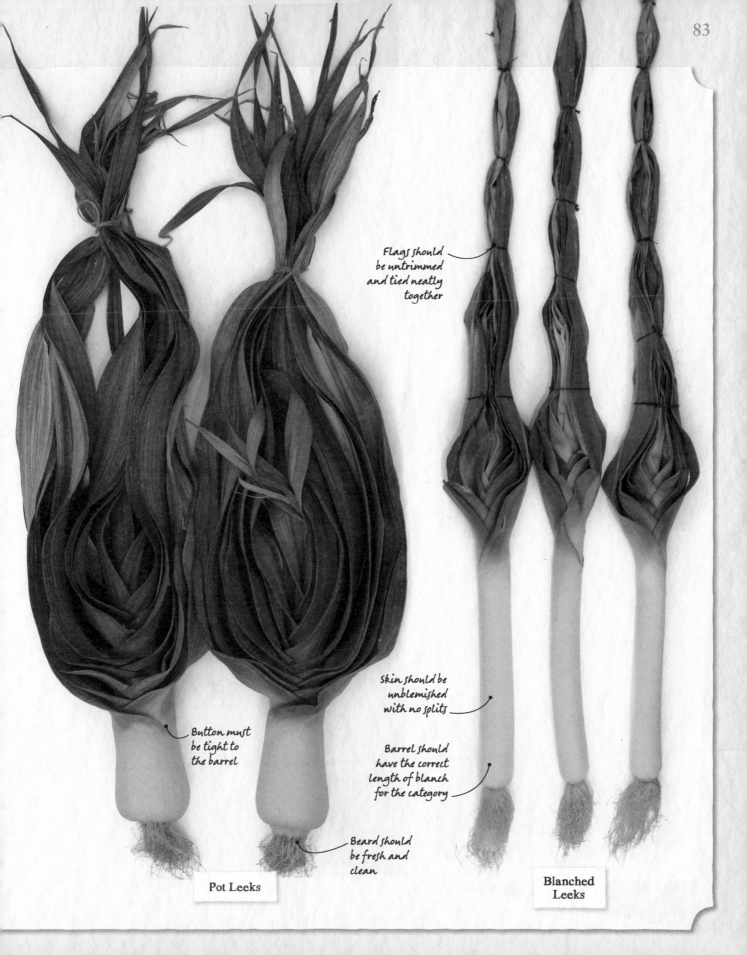

Flags should
be untrimmed
and tied neatly
together

Skin should be
unblemished
with no splits

Button must
be tight to
the barrel

Barrel should
have the correct
length of blanch
for the category

Beard should
be fresh and
clean

Pot Leeks

Blanched
Leeks

Onions

Competition can be fierce among onion-growers, with seasoned competitors being known to jealously guard their secret compost recipes. But don't be put off – onions are easy to grow and there are several classes you can enter.

JUDGES SCORE HIGHLY
Uniform, well-ripened, firm bulbs, with firm, thin necks of good colour, sound root plates, and unbroken skins with no sign of damage or blemish.

DEFECTS TO AVOID
Soft, misshapen, or blemished bulbs with broken skins; soft, thick, or immature necks; unsound root plates; signs of moisture under the skin.

POINTS AVAILABLE

LARGE EXHIBITION
Condition	6
Uniformity	4
Size	5
Shape	3
Colour	2
TOTAL	20

250G (9OZ) AND UNDER
Condition	5
Uniformity	3
Size	2
Shape	3
Colour	2
TOTAL	15

PICKLING
Condition	3
Uniformity	3
Size	2
Shape	2
Colour	2
TOTAL	12

PREP Harvest a few weeks before the show. Wipe with a soft, damp sponge and dry with kitchen towel. Store until show day by standing them upright on a layer of sawdust in a wooden box in a cool, dry location. A day before the show, soften the necks by dipping them in warm water, then tie with raffia and trim to about 20mm (³/₄in). Trim roots back to the basal plate.

PRESENT Large exhibition onions are often displayed raised up on cloth-covered rings (cloth ensures the rings do not mark the bulbs' bases). Entries at 250g (9oz) and under may also be shown on soft collars, or placed directly on to plates with the raffia-tied necks uppermost. Stage pickling onions on a plate with a bed of fine sand for stability.

JUDGING NOTES Schedules normally ask for 3 specimens in the exhibition class, 5 at 250g (9oz) or under, and 12 pickling onions; smaller shows may require fewer onions. All specimens in the 250g (9oz) or under class are weighed; bulbs as close to 250g (9oz) as possible will be preferred by the judges and any over 250g (9oz) will be disqualified. Pickling onions should be approximately 30mm (1in) in diameter. Exhibition onions should always be large.

Undercover Onions
Skin discoloration and marks on the base of your onions can be the difference between winning and losing. Cover the more sensitive skins of exhibition onions with clean J-cloths or dish towels until judging, and if you are placing onions on supports, make sure they are soft enough to avoid marking.

TOP TIPS

To help dry out your onions in time for show day, sprinkle talcum powder over them before placing them in storage. It also helps to turn them on a daily basis, and you could even use an electric fan to circulate air around them.

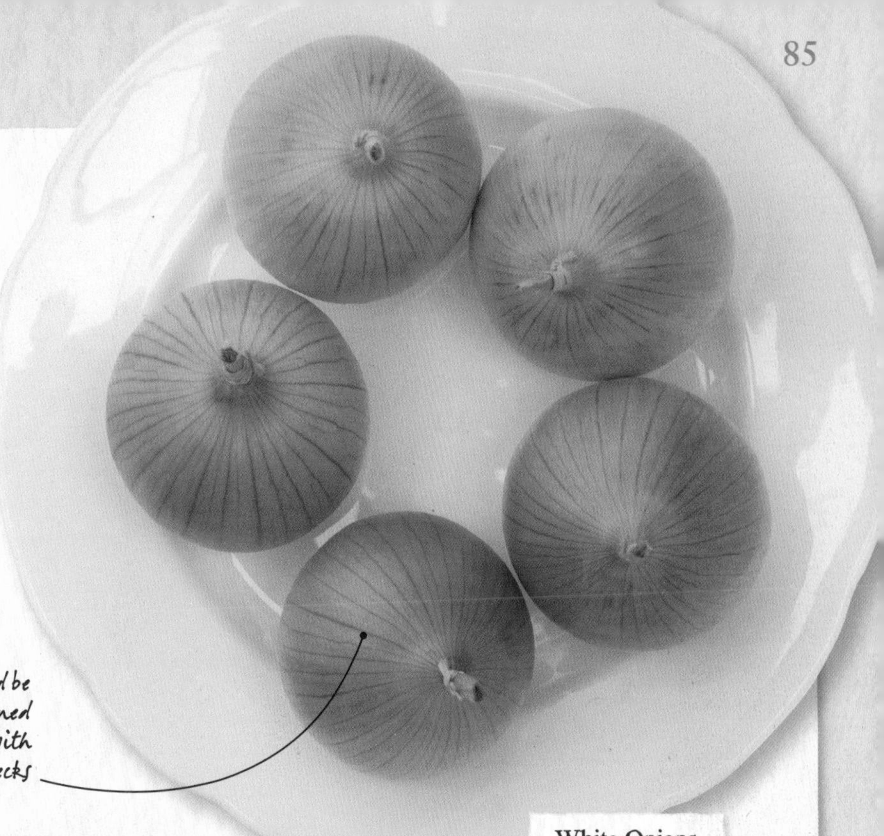

Bulbs should be well-ripened and firm, with thin necks

White Onions, 250g (9oz)

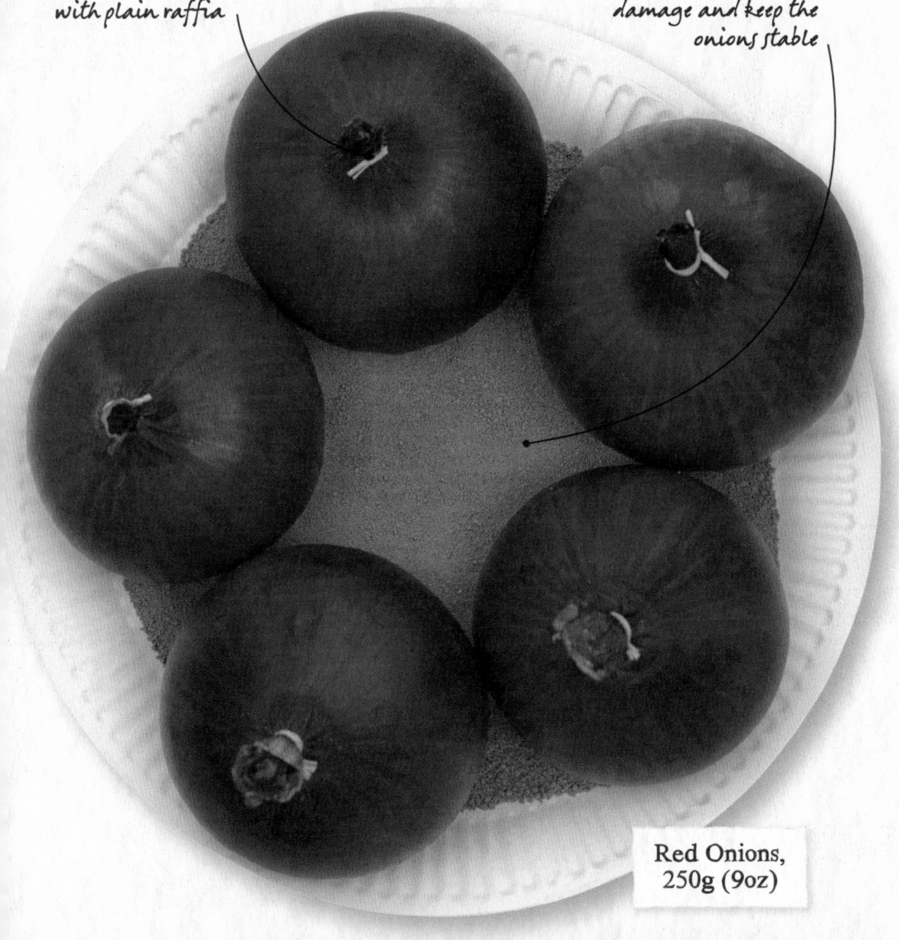

Tie the trimmed necks with plain raffia
Fine sand can help protect the root plate from damage and keep the onions stable

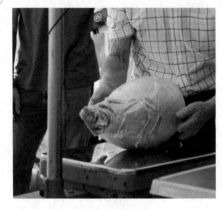

Red Onions, 250g (9oz)

Tipping the Scales

To grow giant onions, you need to artifically extend the growing season. Sow seedlings in late autumn in a heated greenhouse equipped with artificial light so that, by late winter, they are ready to be grown in progressively larger containers housed in a ventilated polytunnel. Carry out a proper soil test to ensure the onions can get all their nutrients from the compost.

Shallots

Shallots grow in clusters, each one jostling for space, and this often leads to odd-shaped specimens. Perfectly round shallots are, as a result, highly prized on the show bench.

JUDGES SCORE HIGHLY
Firm, well-ripened, disease-free bulbs with good colour that are shapely and round in the cross-section, with thin necks neatly tied with plain raffia.

DEFECTS TO AVOID
Soft, poorly ripened bulbs of asymmetrical shape or incorrect size, with split, broken, or blemished skins, or bulbs that have been overskinned. Thick necks!

POINTS AVAILABLE

LARGE EXHIBITION
Condition	6
Uniformity	3
Size	3
Shape	3
Colour	3
TOTAL	18

PICKLING
Condition	5
Uniformity	4
Size	2
Shape	2
Colour	2
TOTAL	15

PREP Separate the clusters and dry the bulbs thoroughly (see Onions pp.84–85 for advice on drying). Choose specimens that are free from staining and do not have loose skins. Don't overskin the bulbs: judges will be able to spot overskinned shallots by any greening or purpling at the base. Trim the roots to the basal plate, shorten the necks, and tie the tops neatly with plain raffia.

PRESENT Shallots are best staged on plates in a bed of sand so that they can be positioned with necks uppermost. Place a layer of fine dry sand or similar material on a plate, preferably of a contrasting colour, and pile the sand slightly in the centre so that the bulbs in the middle of the display are slightly raised.

JUDGING NOTES Competitors are normally asked to show 12 shallots, or a minimum of 6 at smaller shows. Bulbs entered into the pickling class must not exceed 30mm (1in) in diameter. Judges will usually check they can easily pass through a specially designed ring and those that can't will be automatically disqualified.

Bulbs should be large and perfectly round

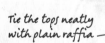

Tie the tops neatly with plain raffia

stage in fine sand for stability

Final Preparations
Take one last look at your shallots after setting them on their sanded plate, to make sure they are well arranged and of a uniform height.

Garlic

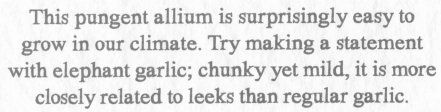

This pungent allium is surprisingly easy to grow in our climate. Try making a statement with elephant garlic; chunky yet mild, it is more closely related to leeks than regular garlic.

PREP Carefully clean the bulbs to remove all soil fragments, taking care not to break the skins, and let the bulbs dry completely (see Onions p.85 for advice on drying). Neatly trim the roots back to the basal plate. For regular garlic, cut the stems to 50mm (2in); for elephant garlic, to 75mm (3in). Keep bulbs whole: don't break them down into cloves.

PRESENT Stage garlic bulbs on plates in a neat arrangement, typically with all stems pointing in the same direction.

JUDGING NOTES Entrants are normally asked to show 5 bulbs of regular garlic varieties, or a minimum of 3 at smaller shows. For elephant garlic, it is 3 bulbs or a minimum of 2.

JUDGES SCORE HIGHLY
Clean, well-ripened, solid, well-shaped bulbs of uniform size with roots removed and skins intact. Large specimens of elephant garlic.

DEFECTS TO AVOID
Bulbs that are misshapen, soft, poorly ripened or immature, with damaged or yellowing skins; bulbs that have split to show the inner segments; regular garlic with thick necks.

POINTS AVAILABLE

REGULAR

Condition	5
Uniformity	3
Size	2
Colour	2
TOTAL	12

ELEPHANT

Condition	5
Uniformity	3
Size	4
Colour	3
TOTAL	15

cut the stems to 75mm (3in)

Arrange bulbs in a neat formation

Make sure the bulbs are clean and the skins unbroken

Roots should be trimmed

cut the stems to 50mm (2in)

Elephant Garlic

Regular Garlic

Bulb ends should be straight, not swollen

Carefully clean the roots to remove any soil particles

Leave roots untrimmed

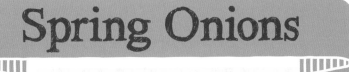

Spring Onions

Also known as green salad onions, spring onions can actually crop as late as September. The tried-and-tested cultivar show bench favourite is 'White Lisbon' – a true classic.

PREP Choose specimens of a uniform size, without swollen bulbs. Lift more than you need in case any are damaged during cleaning or transit. Keep all leaves and roots attached; do not trim. Specimens should be carefully washed before showing.

PRESENT Stage on a plate lined up side-by-side in tight formation, with roots facing the front.

JUDGING NOTES Competitors are normally asked to submit 12 specimens, or a minimum of 6 for smaller shows. Judges will be particularly looking out for uniform plants that are free of pest and disease damage. Red cultivars need not exhibit pure white bases.

JUDGES SCORE HIGHLY
Uniform specimens with fresh, tender leaves and bases that are straight and not swollen.

DEFECTS TO AVOID
Trimmed, damaged, or yellow-tipped leaves. Swollen, bulbous bases that aren't white (except for red cultivars).

POINTS AVAILABLE

Condition	3
Uniformity	3
Size	2
Shape	2
Colour	2
TOTAL	12

Celery

Cultivars of celery can be divided into two classes: "trench" celery is manually blanched (protected from sunlight), while "self-blanching" cultivars are naturally off-white in colour.

PREP Avoid any specimens that have signs of flower heads. Place a tie around the base of the leaves to prevent them breaking. Clean thoroughly, flushing continuously with water to ensure all pests are removed. Trim the roots down to a pointed butt end.

PRESENT Lay entries flat on the show bench and cover them with clean paper or a damp cloth to exclude light; remove any covering before judging starts.

JUDGING NOTES Entrants should submit 2 specimens. Judges will remove ties to check the foliage, then will examine the blanch and stalk arrangement. They will also check the heart for diseases, such as heart rot.

Check leaves for pest damage and any pests hiding within!

Blanch should be evenly distributed

Roots should be trimmed to form a point

JUDGES SCORE HIGHLY
Large, clean, firm, unblemished, well-blanched specimens, with crisp leaf stalks and intact root plates.

DEFECTS TO AVOID
Small, poorly blanched specimens. Evidence of pest damage. Leaf stalks that are thin, twisted, pithy, split, or showing sideshoots. Hearts showing visible flower stalk. Evidence of rot.

POINTS AVAILABLE

TRENCH	
Condition	5
Uniformity	4
Size	4
Shape	4
Colour	3
TOTAL	20

SELF-BLANCHING	
Condition	5
Uniformity	4
Size	3
Shape	3
Colour	3
TOTAL	18

Carte Blanche

Blanching typically produces longer, less stringy, and milder-tasting celery. Growers "collar" their stems by wrapping them in corrugated card when they reach about 30cm (12in) in height, securing the collars loosely in place with twine.

Celeriac

Knobbly and bulbous, even a perfectly grown celeriac will still look rather ugly. That said, the judges do expect quality specimens, so a little bit of care and attention is a must.

JUDGES SCORE HIGHLY
Blemish-free, globe-shaped specimens with correctly trimmed foliage and roots.

DEFECTS TO AVOID
Roots that are rough, split, or flat.

POINTS AVAILABLE

Condition	5
Uniformity	4
Size	3
Shape	3
TOTAL	15

PREP Trim foliage back to the youngest central leaves, leaving them at a length of approximately 75mm (3in). Trim roots back to the body of the vegetable.

PRESENT Specimens may be presented on a plate or laid directly on to the show table.

JUDGING NOTES Entrants are normally asked for 2 specimens of celeriac, or a minimum of 1 at smaller shows. Judges will not consider colouration around the root to be a defect as it is a natural consequence of trimming. Celeriac may be exhibited in both root vegetable and salad classes.

Leave the youngest central leaves intact when trimming foliage

Choose well-rounded, globe-shaped specimens

Thirsty Work

The wild ancestor of celeriac thrives in boggy ground and the cultivated roots also prefer a damp corner and constant supply of water. From midsummer, as they start to swell, remove the outer leaves and any side shoots as they appear, to direct all moisture towards the thirsty bulbs.

Colouration around the roots is not considered a defect

Present large, fleshy bulbs

Trim roots to leave a neat root plate

Trim the foliage of all but the two central stems

Florence Fennel

While related to the herb, Florence fennel is a distinct plant cultivated for its edible "bulbs". It was first grown in England by the Earl of Peterborough, who ate the bulbs as a dessert.

PREP Fennel is sometimes harvested by cutting the bulbs above the root plate, but for exhibiting they should be lifted from the soil whole; the roots are then trimmed but the root plate is left intact. Cut back foliage to 75–100mm (3–4in), but retain the central, "terminal" foliage.

PRESENT Lay bulbs flat, either on a plate or directly on the show bench, with root plates facing the front of the table.

JUDGING NOTES Competitors are normally asked to exhibit 2 specimens, or just 1 at smaller shows. Florence fennel may also be exhibited in a general salad vegetable class, as it can be enjoyed raw in salads. Fennel should not be entered into a general class for root vegetables as it's considered a bulb rather than a root vegetable (though actually the "bulbs" are swollen leaf stems).

JUDGES SCORE HIGHLY
Large, clean, trimmed bulbs with swollen, fleshy leaf bases, lacking coarseness or flower stems.

DEFECTS TO AVOID
Leaves that are small at the base, coarse or loose; bulbs that are flat or elongated; evidence of pest or disease damage.

POINTS AVAILABLE

Condition	4
Uniformity	3
Size	4
Shape	4
TOTAL	15

Judging the Exhibits

The exhibitors clear the tent as
the judges arrive, ready to assess
all the entries in each class.

For each class, judges assess the entries
based on the criteria defined in the show
schedule. Every point they give (or deduct)
must be justified against this set criteria.

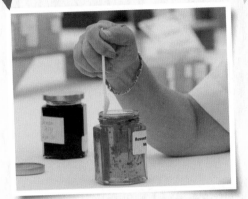

Domestic judges taste every item on the show bench. Large bakes, such as cakes and bread loaves, are often sliced in half to assess the crumb structure within.

Larger shows may have a whole team of judges assessing entries, with each judge specialising in a particular class or category. A post-judging debrief offers them the chance to evaluate the overall standard of entries on show.

Carrots

Growing up to a metre in length, long-pointed carrots often catch the attention of show visitors when laid on the bench. The stocky stump-rooted cultivars may lack the same wow factor, but the judges will inspect them with equal vigour.

JUDGES SCORE HIGHLY

Fresh, firm, long, smooth carrots of a good, uniform shape and weight. Skins should be clean and bright, with no evidence of side roots, and a good colour across the whole carrot. Stump-rooted entries must have a markedly stump end.

DEFECTS TO AVOID

Coarse, misshapen, or split carrots, with dull, pale, poor colouring (or coloured crowns), showing disease or pest damage, multiple roots, or signs of going to seed.

POINTS AVAILABLE

LONG-POINTED

Condition	5
Uniformity	4
Size	4
Shape	3
Colour	4
TOTAL	20

STUMP-ROOTED

Condition	5
Uniformity	4
Size	3
Shape	3
Colour	3
TOTAL	18

PREP Lift carrots very carefully to avoid snapping the fine taproots. Clean by working around the circumference with a wet cloth or sponge, being careful not to remove a layer of skin. Trim the foliage to 50–75mm (2–3in).

PRESENT Lay carrots on a plate, board, or directly on the table, in a triangular formation with the roots facing the front of the table.

JUDGING NOTES Competitors are normally asked to show 3 specimens of either type, or a minimum of 2 at smaller shows. Carrots may also be entered into collection classes for both root and salad vegetables (see p.55). Long-pointed carrots have large, long roots that gradually taper and end in a fine taproot. They are traditionally grown in barrels or deep raised beds, filled mainly with sand. Stump-rooted carrots are shorter and develop a blunt end, or "stump". Chantenay types are broad at the leaf end and slightly tapering, while Nantes types have straighter, more cylindrical roots.

Heavyweight Beasts

Whereas ordinary carrot classes demand fresh, edible specimens with no sideshoots or split ends, in the competitive world of heavyweight vegetables, good looks and flavour play no part and there are no limits to how monstrous, knobbly, or woody a prize-winning carrot can become – it all contributes to the final weigh-in. The current world and UK record for the heaviest carrot is a bicep-ripping 9.09kg (20lb), compared to about 100g (3½oz) for your average common-or-garden carrot.

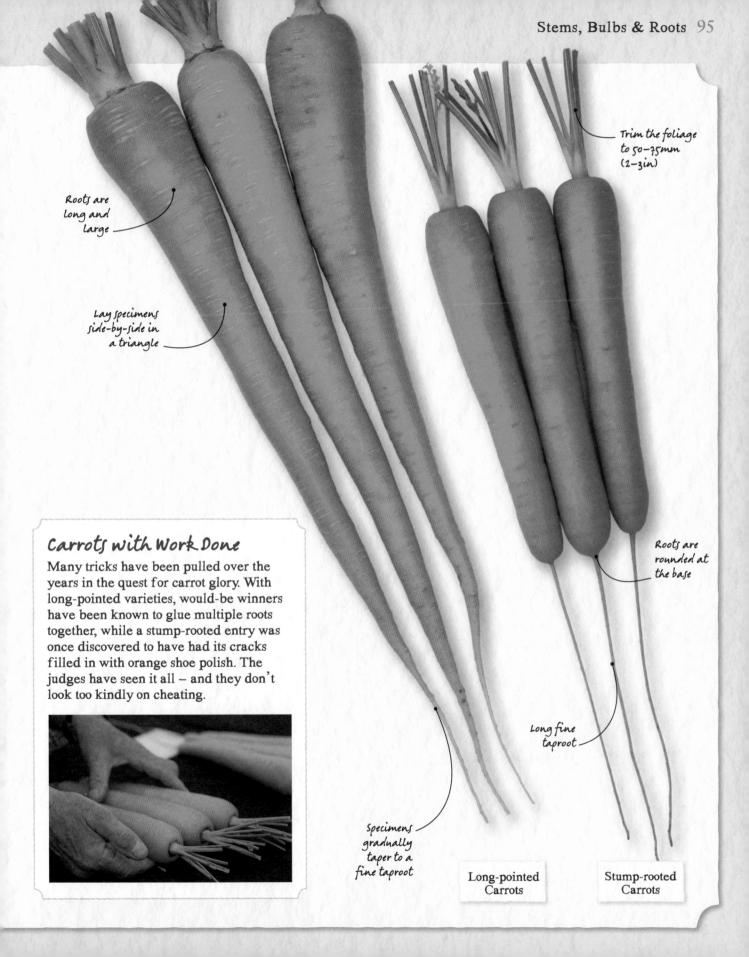

Roots are long and large

Lay specimens side-by-side in a triangle

Trim the foliage to 50–75mm (2–3in)

Roots are rounded at the base

Long fine taproot

Carrots with Work Done

Many tricks have been pulled over the years in the quest for carrot glory. With long-pointed varieties, would-be winners have been known to glue multiple roots together, while a stump-rooted entry was once discovered to have had its cracks filled in with orange shoe polish. The judges have seen it all – and they don't look too kindly on cheating.

Specimens gradually taper to a fine taproot

Long-pointed Carrots

Stump-rooted Carrots

Beetroot

Trouble-free and easy to grow, beetroot is an ideal root vegetable for the village show newcomer. Schedules may offer up to three beetroot classes: cylindrical, globe, and long.

PREP Choose well-sized specimens for the intended class: cylindrical beetroot should be about 150mm (6in) long; globe beetroot should be 60–75mm (2½–3in) in diameter; long beetroot can be any length. Trim foliage to 75mm (3in) and remove any small side roots. Wash away soil carefully so as not to leave any marks.

PRESENT Depending on the size and type, arrange either on a plate or directly on the table. Long cultivars with tapered roots should take a triangular formation. Beetroot may also be entered into salad vegetable classes (see p.55).

JUDGING NOTES Exhibitors must show 4 cylindrical and globe specimens, or 3 for long cultivars.

JUDGES SCORE HIGHLY
Uniform size and colour, with clean, undamaged skin and a single root.

DEFECTS TO AVOID
Over- or undersized, misshapen, or multi-rooted specimens. Rough, corky, damaged, or aged skin; poor skin colour at the base of the root.

POINTS AVAILABLE

CYLINDRICAL AND GLOBE	
Condition	5
Uniformity	3
Size	3
Shape	2
Colour	2
TOTAL	15

LONG	
Condition	5
Uniformity	4
Size	4
Shape	4
Colour	3
TOTAL	20

Cylindrical Beetroot

Long Beetroot Present long beetroot, shown here as part of a collection class, in a triangular formation.

Globe Beetroot

cylindrical roots should be equal in length

clean skin carefully, with a soft cloth

Kohlrabi

With its distinctive, sputnik-like looks, kohlrabi makes an eye-catching exhibit on the show table. This brassica develops fast, with some cultivars maturing from seed in eight weeks.

PREP Cut down the side foliage to about 20cm (8in), but retain the terminal (central) foliage. Trim the roots neatly. Clean, but do not wash, the vegetable as it should retain its natural bloom.

PRESENT Stage specimens on a plate. Kohlrabi can be awkward to arrange as the trimmed leaf bases tend to jut out at different angles, but careful positioning can result in an arrestingly graphic display.

JUDGING NOTES Competitors are normally asked to show 3–5 specimens, or a minimum of 2 at smaller shows. Kohlrabi may be exhibited in both root vegetable and salad vegetable classes.

JUDGES SCORE HIGHLY
Fresh, tender, round swollen stems showing their natural bloom, with small leaf bases and no signs of damage; side foliage trimmed but with the terminal foliage retained.

DEFECTS TO AVOID
Misshapen, cracked, or damaged swollen stems, with a lack of natural bloom and coarse leaf bases.

POINTS AVAILABLE
Condition	5
Uniformity	3
Size	2
Shape	2
TOTAL	12

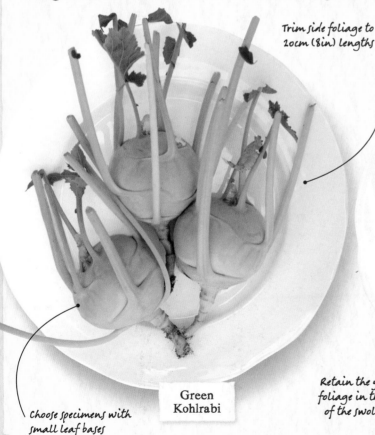

Trim side foliage to 20cm (8in) lengths

Green Kohlrabi

Choose specimens with small leaf bases

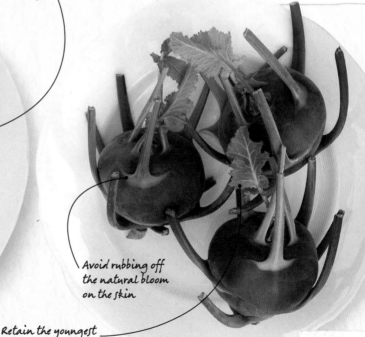

Avoid rubbing off the natural bloom on the skin

Retain the youngest foliage in the centre of the swollen stem

Purple Kohlrabi

Turnips

Turnips are unfairly maligned. While they can turn woody if left to grow too big, picked young they are sweet and spicy. Roots just larger than a golf ball are perfect for both showing and eating.

JUDGES SCORE HIGHLY
Clean, fresh, tender, pest- and disease-free roots, of a shape and size characteristic of the cultivar but not over-large, with a small, single taproot and clear skin.

DEFECTS TO AVOID
Very small or very large roots, with an irregular shape, spongy texture, patchy skins, or signs of multiple taproots.

POINTS AVAILABLE

Condition	5
Uniformity	4
Size	2
Shape	2
Colour	2
TOTAL	15

PREP Before lifting the turnips, soak the soil to loosen it. Lift carefully, keeping the taproot attached. Wash away any soil from the roots and remove any dead foliage. Cut a spare root to check the condition of the insides and that the crop is free from disease. Trim the foliage to 75mm (3in).

PRESENT Stage on a plate, with the taproots facing the front.

JUDGING NOTES Competitors are normally asked to show 3 turnips, or a minimum of 2 at smaller shows. Turnips can be exhibited in either root vegetable or salad collection classes.

TOP TIPS
It's worth growing a second crop of turnips just for the leaves, which are delicious when picked young and lightly steamed. Sow in mid-autumn for turnip tops to help fill the hungry gap from midwinter to early spring.

Skins should be clear, with no sign of pest or disease damage

Size and shape must be typical of the cultivar and not over-large

Specimens should have only one small taproot

Shoulders should be symmetrical

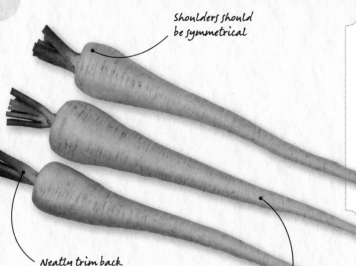

Deep and Long

For top-quality parsnips, you need deep, crumbly soil. Top growers use plastic drums filled with sand to reach the right depth, and make planting holes with a crowbar, which they fill with finely sieved compost.

The taproot is the thin extension of the main body of the parsnip.

Neatly trim back the foliage to about 75mm (3in) long

Parsnips should be well tapered and have smooth, white skins.

Parsnips

Lifting parsnips without breaking the tip is a necessary challenge, as points will be docked for a snapped taproot. Some experts don't lift at all and simply blast the compost away with a hose.

PREP Soak the soil around the parsnips before lifting; this loosens the ground and minimises the risk of root damage. Take great care as bruising by fingers and scratching by soil particles will show up later. Wash the roots thoroughly with clean water and trim the foliage to approximately 75mm (3in).

PRESENT Lay directly on to the show bench, side by side, with the foliage facing away from the front of the bench. Parsnips can also be staged on a blackboard or on a board covered with black cloth.

JUDGING NOTES Entrants are normally asked to show 3 parsnips, or a minimum of 2 at smaller shows. Parsnips may also be entered into collections of root vegetables. Although long specimens are preferred, good weight is considered more important than length.

JUDGES SCORE HIGHLY
Large, long, uniform roots with good heft, shapely and evenly tapered, with smooth, unblemished skins.

DEFECTS TO AVOID
Small, dirty roots with side roots present; rough, discoloured skins; misshapen shoulders. Side roots, blemishes, canker, or excessive ribbing. Broken or missing taproot.

POINTS AVAILABLE

Condition	5
Uniformity	4
Size	4
Shape	4
Colour	3
TOTAL	20

Jerusalem Artichokes

Jerusalem artichokes are one of the simplest vegetables to grow and a good choice if you're new to showing, but you'll need to find a late autumn or early spring show for this wintery crop.

JUDGES SCORE HIGHLY
Large, shapely tubers with smooth, unblemished skins, of a colour true to the cultivar.

DEFECTS TO AVOID
Small, oversized, irregularly shaped tubers with damaged, patchy or rough skins.

POINTS AVAILABLE

Condition	4
Uniformity	2
Size	2
Shape	2
TOTAL	10

PREP Lift tubers with a fork, being careful not to pierce any of them. Use a sharp knife to cut the tubers from the root clump. Select the least knobbly and most uniform-looking specimens, and trim the stem ends. Clean with a cloth and warm water to remove any soil, but do not scrub so that you damage the skin.

PRESENT Stage on a plate in a circular formation, with the cut ends pointing inwards.

JUDGING NOTES Entrants are normally asked to submit 6 tubers, or a minimum of 3 at smaller shows. Judges may cut open one tuber to check freshness and inspect the internal condition.

Tubers should be relatively uniform in shape and size

Skins should be clean and undamaged

Choose the largest tubers from your crop

Reaching for the Sun

These tuberous root vegetables are unrelated to globe artichokes and are in fact part of the sunflower family. The Jerusalem element of their name derives from the Italian for sunflower: "girasole". Like lots of sunflowers, they grow very tall and may need staking to stop them blowing over in strong winds.

Potatoes

Grown for show, the humble potato can be a beautiful object, especially those in the coloured class, which can include fully coloured cultivars or those with coloured "eyes" (dormant buds).

PREP Lift with a fork, trying not to pierce any, and select medium-sized specimens for the cultivar, which should generally weigh 200–250g (7–9oz). Very carefully wash the tubers with water and a soft sponge, but do not damage the skins by scrubbing with a brush.

PRESENT Stage on plates with the "rose" ends (where the eyes are concentrated) facing outwards. Cover with a cloth to exclude light, removing it just before judging.

JUDGING NOTES Entrants must normally submit 5 specimens, or a minimum of 3 at smaller shows. Schedules may include separate classes for white and coloured potatoes, though they are judged by the same criteria. Competitors with a normally coloured cultivar that shows no colour may be able to enter them in the white class; check with the organisers if you are uncertain. Salad potatoes may be entered into a collection class for salad vegetables (see p.55).

JUDGES SCORE HIGHLY
Shapely, clean, medium-sized tubers with clear skins and shallow eyes, well-coloured appropriate to the cultivar.

DEFECTS TO AVOID
Tubers that are too small or large, damaged or misshapen, with patchy or speckled skins and excessively deep eyes; any signs of greening.

POINTS AVAILABLE
Condition	5
Uniformity	5
Size	3
Shape	4
Eyes	3
TOTAL	20

TOP TIP
If you live in an area where the tap water is "hard" (has a high mineral content), it is best to water your potato crop with rainwater to achieve blemish-free skin, as hard water is known to encourage common potato scab.

Choose the largest tubers from your crop

Tubers should be relatively uniform in shape and size

Position with the "rose" ends pointing out

Radishes

The familiar small, round, red salad radishes are still the most common types of radish to be found on the show bench, but the much larger Oriental and winter types are beginning to make their more substantial presences felt.

JUDGES SCORE HIGHLY

Uniform, young, fresh, and firm but tender specimens that are well-coloured and blemish-free, of a size characteristic of the cultivar. Salad radishes with foliage that is fresh, clean, and correctly trimmed. Oriental and winter radishes with clean but untrimmed leaves.

DEFECTS TO AVOID

Old, tough, misshapen, limp or spongy roots, of a dull colour, that show pest or disease damage or are running to seed. Salad radishes with foliage that is untrimmed or trimmed to a length that is substantially more or less than 30mm (1in).

POINTS AVAILABLE

SALAD

Condition	4
Uniformity	3
Size	3
Colour	2
TOTAL	12

ORIENTAL AND WINTER

Condition	5
Uniformity	4
Size	3
Colour	3
TOTAL	15

PREP Dig up radishes close to show time, so that they retain maximum turgidity. Select young specimens that haven't grown too big; cut a few spare to check the internal condition. Take care not to snap off the fine taproots. Trim the foliage of salad radishes to approximately 30mm (1in), but leave the foliage of the larger Oriental and winter radishes untrimmed. Wash specimens to remove any soil particles.

PRESENT Salad radishes are best staged in a circle on a plate, with the trimmed leaf stalks pointing inwards. Depending on size, Oriental and winter radishes may be presented directly on the show bench, with roots to the front, or arranged on plates.

JUDGING NOTES Entrants must normally submit 9 salad radishes, or a minimum of 5 at smaller shows; 3 Oriental or winter types, or a minimum of 2. Salad radishes are defined as quick-growing and tender-rooted, and include three types: round; cylindrical French breakfast cultivars with blunt ends; and white-pointed. Oriental and winter radishes are defined as large, round, or long-pointed roots, weighing up to 2kg (4lb) in the case of Oriental radishes and up to 1kg (2¼lb) for winter radishes.

Knot Easy

Oriental radishes grown for the heaviest class gain weight by developing multiple side roots like knotted, contorted limbs, giving them a characterful, almost tortured look. The world record for heaviest radish is a whopping 31kg (68lb).

Gently Does It

Staging entries to the longest radish class requires true patience and a very steady hand to uncoil the full length of the delicate taproot, which contributes most to the final measurement. The world record stands at just over 5m (16ft).

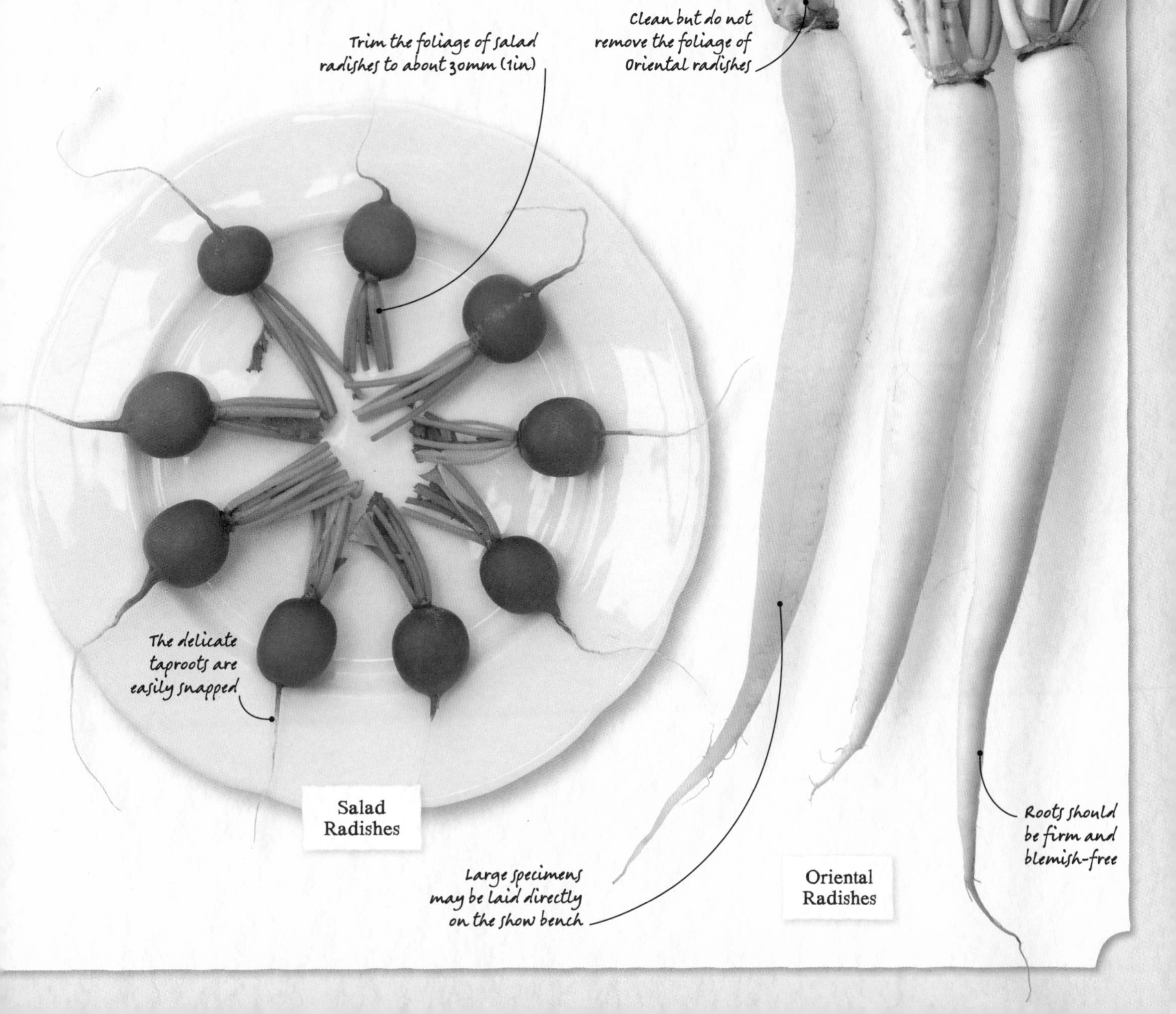

Trim the foliage of salad radishes to about 30mm (1in)

Clean but do not remove the foliage of Oriental radishes

The delicate taproots are easily snapped

Salad Radishes

Large specimens may be laid directly on the show bench

Oriental Radishes

Roots should be firm and blemish-free

Lettuces

For ancient Egyptians lettuce was an aphrodisiac; for the Romans it promoted chastity; for us it is the basis of a good salad. The main issue on the show bench is how to stop your specimens going limp, as only the firmest heads win prizes.

JUDGES SCORE HIGHLY

Fresh, clean, tender, unbroken, and blemish-free heads of a good, uniform size and colour appropriate to the cultivar, with the roots intact. The leaves of cos and crisphead lettuces should be crisp.

DEFECTS TO AVOID

Dirty, over-trimmed, limp heads that are blemished or a poor colour; signs of bolting (production of flower heads) and damage from pests or diseases; absence of roots.

POINTS AVAILABLE

CRISPHEAD, COS, AND BUTTERHEAD

Condition	6
Uniformity	4
Firmness and texture	3
Colour	2
TOTAL	15

LOOSE-LEAF

Condition	5
Uniformity	4
Colour	3
TOTAL	12

PREP Lift the lettuces, with their roots intact, the evening before show day or early the same morning, when the leaves are fresh and crisp. Remove any damaged outside leaves. Wash the roots and the heads gently, dislodging any soil particles that may have collected between the leaves. Wrap the roots in moist tissue, insert them into a plastic bag, and neatly tie the bag so that it is firmly attached to the roots.

PRESENT Lay the heads on the show bench, with the hearts facing the front of the bench. Leave the roots wrapped and the plastic bag attached for the duration of the show.

JUDGING NOTES Exhibitors are normally asked to show 2 heads of lettuce at both large and small shows. Judges may open the bag containing the roots to check they are present. There are two types of lettuce: those that form relatively solid heads of leaves and loose-leaf varieties that do not. Within the head-forming category, there are crispheads with tightly folded, crunchy leaves and a solid heart; butterheads with soft, folded leaves and round hearts; and Cos types that are more upright and oval-shaped, with looser hearts. Head-forming and loose-leaf lettuces are normally distinct classes in a schedule and judged separately.

Cool Customers

Lettuce seeds like it cool and seed germination is poor if soil temperatures go above 25°C (77°F). The critical period is a few hours after planting, so during hot weather it is best to sow seeds in the late afternoon or evening, when temperatures will be on a steady decline. Watering rows before you sow will also help to keep seeds cool, as does artificial shading for rows both before and after sowing.

Crisphead
Lettuces

Crisphead
leaves should
have a crisp
texture

Leaves should
be unbroken and
blemish-free

Butterhead
Lettuces

carefully clean
the leaves so that
they are free of
soil particles

TOP TIP

Water lettuce plants well and
regularly, especially in periods
of hot weather and drought.
Inadequate watering places plants
under stress and forces them
to "bolt", which is when they
put energy into producing
flowers rather than
leaves.

Butterheads
have rounded
hearts

Cabbages

Cabbages may no longer be the kings of the brassica world thanks to upstart kale, but they still cut impressive figures on the show bench where you'll find classes for green, red, and Savoy cultivars. All are judged by the same criteria.

PREP Trim cabbages with a sharp knife to leave a stalk about 75mm (3in) long. Gently brush away any soil. Handle carefully so as not to mark the bloom and remove only a minimum of outer leaves.

PRESENT Stage directly on to the show bench with the stalks facing away from the front.

JUDGING NOTES Competitors are normally asked to enter 2 specimens of the same cabbage cultivar, or 1 specimen at smaller village shows. Cabbages are classed as salad vegetables for show purposes and, as well as the individual class, may be entered in a general salad class or as part of a salad collection.

JUDGES SCORE HIGHLY
Fresh, clean, dense heads of equal size, with a waxy bloom, and leaves free from any pest or disease damage. Colour and shape should be characteristic of the cultivar.

DEFECTS TO AVOID
Soft, split heads that are not fresh, with signs of pest and disease damage on the leaves, or with too many outer leaves removed. A stalk that is too short or missing.

POINTS AVAILABLE

Condition	5
Uniformity	3
Size	3
Shape	2
Colour	2
TOTAL	15

The cabbage is a village show constant – no matter what time of year, one cultivar group will be reaching maturity.

Thwarting the Enemy

To grow show-quality cabbages you will need to keep a few beasties at bay. Among their enemies are cabbage root fly maggots, the caterpillars of the cabbage white butterfly, and birds (pigeons especially). Brassica collars, bought or made from pieces of carpet underlay, stop the female cabbage root fly laying eggs around the stems of seedlings, while netting or fleece acts as a barrier against birds and the egg-laying ambitions of cabbage white butterflies.

Fly Collar
Fit collars immediately after planting your seedlings to guard against the attentions of cabbage root flies.

Netting the Crop
The mesh of the netting or fleece must be fine enough to prevent beaks and butterflies penetrating through.

Big Heads

Cabbages measuring over a metre wide are common in the giant veg competition. The world record for a green cabbage stands at nearly 63kg (139lb). Red varieties are positively petite by comparison, weighing in at just over 23kg (51lb).

Heads should be dense with the bloom intact

Select specimens uniform in shape and size

Red Cabbages

Show specimens with as many outer leaves as possible

Savoy cabbages have deeply veined, crinkly leaves

Savoy Cabbages

The Horticultural Society Chairman

My name's Tony Pizzoferro and I am chairman of the Lambeth Horticultural Society. I help to organise our two shows each year, in April and September.

STARTING OUT

I first became involved in the Society when I was a teenager. I joined at Lambeth's inaugural show, back in 1974. Before then, the Society held flower competitions in a small marquee in the park. Then the council came along, wanting to expand the competition and create a bigger, more diverse event. And so the Lambeth Country Show was born.

Since then, the event has been held almost every year, and interest in the competitions just keeps on growing. At our most recent show, we had more than 100 people enter and exhibit, in every class except one: gladioli – it's been a bad year for gladioli...

Now that I'm involved in organising the shows, I don't really exhibit any more. I used to enter something into most of the main categories, but my real love lies with cacti and succulents.

LIFECYCLE OF A SHOW BENCH

You see the popularity of different produce wax and wane from one year to the next. People might see such-and-such win a certain class and think they'll never have a chance against them. Because of that, they don't try entering the class themselves, which of course makes it easier for that winning competitor to keep their title

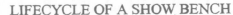

Experts on Tap

Volunteers from Tony's Society man the benches to answer any questions and hopefully inspire visitors to enter the next year.

Sharing the Knowledge
Neat staging is vital, says Tony, and experienced exhibitors will be glad to explain the tricks of the trade.

"Keep an eye out for calyces – they're the husks and a sign that something's home-grown. We've had cheaters in the past stick calyces on to supermarket veg in an attempt to win!"

year after year. Finally, someone does have a go and manages to beat the reigning champion, which opens the field again for more new talent to try competing. And so the cycle goes on.

ADVICE TO BEGINNERS

If you want to start exhibiting, I always say you should speak first to people who've entered before. You'll get advice on all sorts of things, like how best to dress your onions and trim your beetroots, how to transport delicate berries, and so on. Even though it's a competition, I've always found gardeners to be people who want to share their expertise and knowledge.

It's also good to visit the show you want to enter before you start showing your own exhibits there. It's long-term preparation, but gives you an idea of what to expect and if you're at

the same level. Most importantly, just get involved. You never know if you can win until you enter. Even if you don't succeed, there's enjoyment out of seeing everyone's efforts. That camaderie you see between competitors really is the best part of showing.

Neat Ideas
Tony recommends visiting shows to pick up new ideas for staging entries, like tying the stalks of beetroot leaves with plain raffia.

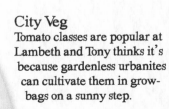

City Veg
Tomato classes are popular at Lambeth and Tony thinks it's because gardenless urbanites can cultivate them in grow-bags on a sunny step.

Kale

Kale has shot to fame recently as a superstar amongst superfoods. With purple-stemmed and red-leafed cultivars available, this brassica makes an attractive exhibit at late shows.

JUDGES SCORE HIGHLY
Fresh, clean, well-developed, and blemish-free leaves, of a good colour and size characteristic of the cultivar.

DEFECTS TO AVOID
Limp, poorly developed leaves that are damaged or of a poor colour.

POINTS AVAILABLE

Condition	5
Uniformity	3
Size	2
Colour	2
TOTAL	12

PREP Pull fresh, young but well-developed leaves from the centre of plants as close to show time as possible. Gently wash the leaves to remove any soil.

PRESENT Exhibit leaves in a glass vase filled with water, to maintain freshness. The vase should be plain and transparent so that the full length of the stems may be examined.

JUDGING NOTES Exhibitors are normally asked to show 10 kale leaves or a minimum of 5 leaves at smaller shows. All types are judged in the same class.

Leaves should have a good colour for the cultivar

Leaves should be fresh, with no holes from pest damage

Stems and leaves should be fully turgid with no signs of limpness

Make sure the vase is stable enough to hold the large leaves

Redbor Kale

Cavolo Nero (Black Kale)

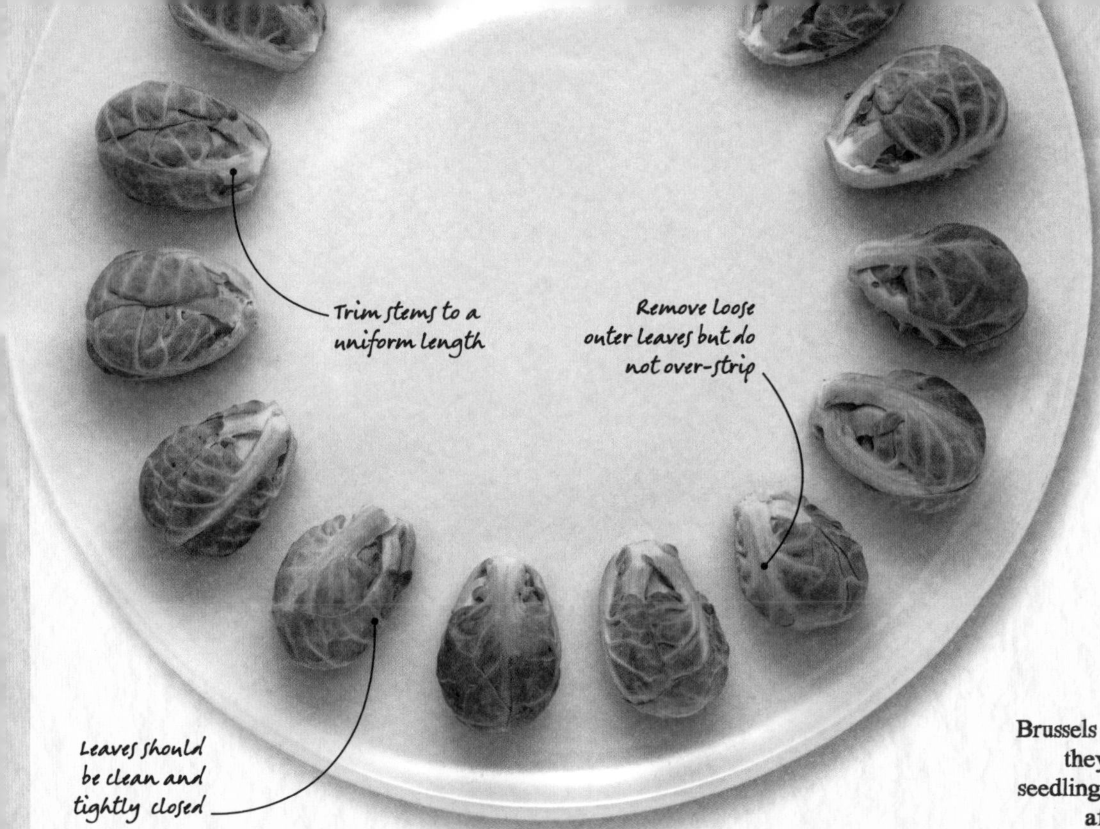

Trim stems to a
uniform length

Remove loose
outer leaves but do
not over-strip

Leaves should
be clean and
tightly closed

Standing Firm
Brussels sprouts have shallow roots but
they grow tall and heavy, so plant
seedlings firmly and earth up or stake
after a few weeks if necessary.

Brussels Sprouts

Brussels tend to be thought of as a midwinter
vegetable, but crops can be harvested as early
as September and you will often find a class
for sprouts in competition at autumn shows.

PREP To encourage even
development of sprouts along
the stems, remove the tops of the
plants when the lowest sprouts
are 1cm (¹/₂in) in diameter. Cut
sprouts from the main stem with
a sharp knife, choosing tightly
closed sprouts of uniform size.
Trim the stalks to roughly the
same length. Remove some outer
leaves, but not too many, as this
diminishes the depth of colour.

PRESENT Stage sprouts on a
plate, arranged in a neat circle
around the edge with the stalk
ends facing the centre.

JUDGING NOTES Competitors
are normally asked to show 15
specimens, or a minimum of 8
at smaller shows.

JUDGES SCORE HIGHLY
Clean, fresh, solid sprouts, with
tightly closed buttons (the points of
overlap between leaves) that are
blemish-free and a good colour.

DEFECTS TO AVOID
Old, yellowing sprouts; loose leaves
or excessive peeling of outer leaves;
signs of pest or disease damage.

POINTS AVAILABLE

Condition	5
Uniformity	4
Size	3
Colour	3
TOTAL	15

Tightly packed florets

Showing 6 florets is permissible at smaller shows

Trim stalks to 75–100mm (3–4in)

Supported Growth

Broccoli plants have shallow roots and often become top-heavy. To keep plants stable, grow them in a bed of firm soil, "earthing up" (mounding the soil) around stems as they grow, and tie them to stakes for additional support against strong winds.

Sprouting Broccoli

Dubbed "Italian asparagus" by Brits when first grown in this country, sprouting broccoli differs from its calabrese cousin in that it grows many small florets rather than one large head.

PREP Select the best specimens, uniform in size and condition, and cut them from the plant using a sharp knife. Trim the florets to leave approximately 75–100mm (3–4in) of stalk remaining. Leaves should be roughly level with the bottom of the florets, so trim off any that are over-long.

PRESENT Stage on a plate, tightly packed side-by-side with stems to the front, or arrayed in a circle, with the flower ends facing out. Cover white florets with paper or cloth to exclude light, but remove immediately before judging.

JUDGING NOTES Competitors are normally asked to submit 12 florets, or a minimum of 6 at smaller shows. White and purple cultivars, and the green sideshoots of calabrese (see opposite), are usually judged in the same class and by the same criteria.

JUDGES SCORE HIGHLY
Uniform, firm, fresh florets with tight heads and a good colour appropriate to the cultivar; florets with the correct length of stem.

DEFECTS TO AVOID
Limp, old florets with spongy texture and poor colour; signs that the florets have "blown" (flowers starting to open).

POINTS AVAILABLE
Condition	5
Uniformity	4
Size	3
Colour	3
TOTAL	15

Calabrese

Calabrese is a fancy name for the type of broccoli we're probably all most familiar with, which forms a large, central head. Sideshoots can be entered into a sprouting broccoli class.

PREP Select the largest, firmest heads with flower buds tightly shut. Cut from the plant with a sharp knife, leaving a length of stalk 75–100mm (3–4in). The specimens should be uniform, and should not be damaged, discoloured, split, or loose.

PRESENT Stage directly on the table with stalks facing the front. Just prior to staging, trim back leaves so that they match the level of the outside of the heads.

JUDGING NOTES Competitors should submit 2 heads of broccoli, or 1 at smaller shows. Romanesco broccoli, with its unusual spiralling, fractal-patterned head, is a type of calabrese (despite also being known as "Roman cauliflower") and must be entered into this class.

JUDGES SCORE HIGHLY
Fresh, solid, tightly closed heads of good colour.

DEFECTS TO AVOID
Heads that lack freshness, are becoming loose, soft, or beginning to flower. Specimens of poor colour or that are turning yellow.

POINTS AVAILABLE

Condition	5
Uniformity	3
Size	3
Shape	2
Colour	2
TOTAL	15

TOP TIPS

Calabrese does not transplant well and is better started off in modules or sown in situ. Plant in moderately fertile soil, but don't overfeed with nitrogen, which encourages soft, leafy growth that renders plants prone to wind and insect damage.

Choose large heads that are firm and fresh

Trim leaves back to be level with the outside of the head

Cut the leaves so that they are level with the outer edge of the curds

Curds should be creamy white and smooth-textured

Scorch Protection
Strong sunlight can scorch cauliflower curds. Wrap heads in their own leaves for protection and loosely tie with string.

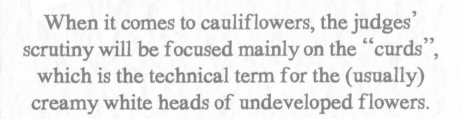

Cauliflowers

When it comes to cauliflowers, the judges' scrutiny will be focused mainly on the "curds", which is the technical term for the (usually) creamy white heads of undeveloped flowers.

PREP Cut cauliflowers to leave a stalk of about 75mm (3in). Select heads with curds that are circular from above and dome-shaped from the side. Avoid specimens where there are visible bracts (small green leaves) between the curd segments. Just before showing, neatly trim back the leaves so that they are level with the surface of the curds. Cover curds with a clean cloth or tissue to prevent discoloration from exposure to the light, but remove it immediately before the judging starts.

PRESENT Stage cauliflowers directly on to the show bench with the curds facing the front.

JUDGING NOTES Competitors are normally asked for 2 heads of cauliflower, or 1 head at smaller shows. Cultivars with green or purple curds have been developed in recent years and these should be classed and judged separately from varieties with white curds.

JUDGES SCORE HIGHLY
Compact, firm, smooth-textured curds, free from blemishes and pest and disease damage; symmetrical, circular, and domed heads; curd colour characteristic of the cultivar.

DEFECTS TO AVOID
Spongy, granular, noticeably lumpy, or loose curds. Under-developed, flat, or asymmetrical heads, with visible bracts, discoloration, or any signs of pest or disease damage.

POINTS AVAILABLE

Condition	5
Uniformity	4
Size	4
Shape	4
Colour	3
TOTAL	20

Chard

With its deep green or bronzed foliage and often vividly coloured stalks, chard is one vegetable that wouldn't look out of place in the ornamental section of the horticultural tent.

PREP Gather leaves at the last moment before the show so that they are as fresh as possible and remain turgid (firm) for judging. Cut complete leaves from the plant with a sharp knife, leaving the stalk intact. Trim the base of each stalk but retain as much of the length as you can, since the stalks are a key feature of the vegetable. Take great care handling the leaves as they can be easily torn.

JUDGING NOTES Schedules normally ask for 15 leaves in an entry, or a minimum of 8 at smaller shows. Leaves and stalks of mixed colours are allowed if they are from a mixed cultivar, such as 'Rainbow'. Spinach beet is a close relative of chard and judged to the same criteria.

PRESENT Display the leaves attractively in a glass vase filled with water, to maintain their freshness. A long, clear vase will show off the stems to good effect.

JUDGES SCORE HIGHLY
A pleasing display of broad, very fresh, undamaged leaves and stalks, of a good colour characteristic of the cultivar, and with no signs of pest or disease damage.

DEFECTS TO AVOID
Limp, torn leaves and broken stems of poor colour, with evidence of damage from pests or disease.

POINTS AVAILABLE

Condition	5
Uniformity	3
Size	2
Colour	2
TOTAL	12

Chard and beetroot share the same species, but varieties of chard have been bred for their leaves rather than roots.

Arrange the leaves so that the stalks are on prominent display

Leaves should be broad and undamaged

Arrange the leaves to create an attractive "bouquet"

Choose a clear, wide, long-necked vase

Herbs

Fresh herbs transform our cooking with their flavour and fragrance, and most are resilient plants that ask for little in return. Herb growers can enter classes for cut herbs displayed in vases or for herbs growing in pots.

JUDGES SCORE HIGHLY
Clean foliage that is fresh, healthy, and blemish-free. Pot-grown herbs should also be sturdy and shapely, and in proportion to the size of pot.

DEFECTS TO AVOID
Unhealthy, yellowing foliage that is showing signs of ageing or pest and disease damage. Pot-grown herbs that look drawn or undernourished.

POINTS AVAILABLE

Condition	6
Size	3
Colour	3
TOTAL	12

PREP Give pots of herbs a quarter- to half-turn every few days to ensure all-round growth. Cut bunches of herbs as close to show time as possible. Handle the herbs carefully, especially those with more delicate leaves, such as basil and tarragon, which bruise easily, and cut extra stems in case any are damaged in transit. Pick leaves clean of soil particles. Remove any foliage below the water level of the vase.

PRESENT Display bunches of single herbs in a vase of water, about 150mm (6in) high and 65mm (2½in) at the mouth, to support the stems. Pot-grown herbs should be shown in pots that are in proportion to the size of the plant.

JUDGING NOTES Competitors must normally show 1 bunch of herbs, and 1 pot if entering a container-grown herb. Schedules will specify the number of herbs to include in a collection. Herbs exhibited in the vegetable section should be for culinary purposes and therefore foliage is the main factor. Seed providers, such as coriander and dill, may be shown for leaves but not for seeds; root providers, such as Florence fennel, should not be included.

TOP TIPS
Do not over-feed your herbs as this produces soft growth that is more prone to damage. Look out for pests, especially rosemary beetle on rosemary, sage, and thyme, and scale insects on bay. Water bay with Epsom salts to prevent yellowing.

Success Lies in Soil
When growing herbs in pots it is vital to get the compost right. Drought-tolerant Mediterranean herbs such as sage, thyme, rosemary, oregano, and bay prefer a free-draining mix of 50:50 loam-based compost and horticultural grit. Plant more thirsty herbs like parsley and mint in fully loam-based compost to retain moisture.

Make sure the Latin names are accurate if included on labels

Select sprigs with clean, unblemished foliage

Apple Mint

SAGE (SALVIA OFFICINALIS)

ROSEMARY (ROSMARINUS OFFICINALIS)

THYME

PARSLEY

CURRY PLANT

Herb Collection

Points are awarded for an attractive arrangement

Handle leaves gently to avoid bruising

Rosemary

Lemon Balm

Schedules may specify showing herb bunches in bikini-style vases

Sage

GROW TO SHOW
FLOWERS

Single stems or blooming bouquets, exotic orchids or eye-catching dahlias: discover the glorious variety of floral classes and how to achieve competition success.

Tips & Tricks
FLOWER CLASSES

The classic image of a flower competition is one of pristine cut flowers in stunning floral arrangements or single-stem exhibits. Most shows will also include several classes for plants in pots, some of which will have been grown indoors. While they share many growing needs, each type of exhibit requires different techniques to get them looking their best on the show bench.

Plants in Pots

PREPARATION To ensure strong "all-round effect" (see All-round Success, opposite) turn pots every few days in the run-up to show day, so that plants are in good condition on all sides. If necessary, gently tease out flowers and leaves for the best display. Remove any damaged foliage or flowers before staging entries.

PRESENTATION Examine the plant from every angle. Remove any foliage and flowers damaged in transit, and ensure the pot is clean. Stage the plant so that its best side is towards the front. Place labels directly beside the exhibit.

TRANSIT Methods of transporting will vary depending on the size and shape of both the plant and its pot. Place the entry in a strong, suitably sized container, strapping the pot into position and protecting any delicate stems with cane supports, if required, to ensure the plant is carried upright and is not crushed or damaged.

Choice of pot is important: it must be in proportion to the size of plant and should not detract from its beauty

Cut Stems

PREPARATION The ideal time to cut flowers is in the evening or early morning – do not cut at midday, when they become dehydrated by strong sunlight. Cut flowers with as much stem as possible, making a slanting cut to maximise water uptake. For flowers that do not absorb water easily, such as chrysanthemums, cut a 75mm (3in) slit up the stem to assist them. Remove sideshoots, unwanted bulbs, and lower leaves, then place the flowers upright in a container of deep, clean water and leave overnight (if possible) in a cool, dark room.

TRANSIT Many exhibitors pack their flowers inserted into bottles in milk crates, and often fix supports to each stem and add packing between flowers to minimise the amount of movement. If using supports for flower stems, they should be removed prior to staging. Always take spare flowers and vases in case of damage in transit.

PRESENTATION Carefully unpack each flower and place in a spare container. Stage in fresh water or water-retaining material, such as "oasis". If using oasis, this should be positioned in the vase as early as is convenient. Stage each vase carefully, ensuring that the stem is neither too long nor too short and that any damaged leaves are removed. Blooms should all face the same direction, unless the schedule indicates "all-round effect" is part of the judging criteria. Place labels directly beside the exhibit; these can be written in advance to save time.

Mixed vases and flower arrangements have unique requirements in addition to the advice given here: see page 126 for more information

All-round Success

"All-round effect" is a desirable quality and a key criterion of judging for most pot plant, mixed vases of cut flowers, and many floral arrangements. In cut-flower exhibits, all-round effect means blooms and foliage should be evenly distributed across the whole display for a balanced, symmetrical presentation. For pot plants grown indoors, it means that there should be an even distribution and quality of growth, and to achieve this the pot must be regularly turned so that all sides face the main source of sunlight for an equal amount of time.

Floral Displays

Floral display classes offer a chance to share your creative talent for flower arranging. Shows often include classes for general mixed vase displays as well as themed arrangements.

JUDGES SCORE HIGHLY
Mixed vases showing an attractive, balanced, and harmonious overall design, with a high proportion of fully developed stems in good, fresh condition and typical of the species or cultivar. Arrangements showing beauty of form and colour, lightness of design, happy harmonies or suitable contrasts, with all plant material in top condition; designs that creatively engage with a theme and show originality of interpretation.

DEFECTS TO AVOID
Mixed vases with foliage in poor condition and flowers that are misshapen, undeveloped, atypical or past their peak; any foliage not sourced from the flowers displayed. Arrangements with unbalanced designs, lacking in rhythm and proportion; plant material in poor condition; designs showing little or no engagement with a theme.

POINTS AVAILABLE
The RHS does not recommend a points scoring system for flower arrangements (see Judging Notes).

MIXED VASES

Condition of flowers, foliage, and stems	8
Colour, texture, and arrangement	6
Symmetry and balance of exhibit presentation	6
TOTAL	20

PREP Condition the flowers by stripping leaves from the bottom half of each stem. Re-cut stems at the base, using secateurs or a sharp knife to ensure a clean cut. Give the stems a good drink by storing them in clean water, somewhere cool, for several hours or overnight.

PRESENT Packing material and discreet supports may be used to help position flowers and foliage. Create the display or arrangement with a "best side" in mind, and ensure this side faces the front of the bench. Once finished, check all is tidy around your exhibit.

JUDGING NOTES In a mixed vase class, do not use foliage from plants other than those of the flowers being exhibited. It should be clear from the schedule if an arrangement class calls for a non-specific or thematic design. Judges will not be influenced by the rarity and cost of the flowers. No points system is recommended by the RHS; the schedule should indicate if the rules of the National Association of Flower Arrangement Societies (NAFAS; www.nafas.org.uk) will apply.

Design should be well-balanced, with pleasing contrasts

House Plants

House plants brighten the home and allow gardenless growers to flex their green fingers. The information here covers all indoor pot plants without a separate entry in the chapter.

PREP Most house plants are judged for "all-round effect" and should be turned every few days to ensure all sides receive enough sunlight and look their best. Use a soft brush to dust very gently both flowers and foliage. Remove any water spots and other marks from foliage with a damp sponge. Make sure the pot is clean, undamaged, and in proportion to the plant.

PRESENT Position the pot so that the best view of the plant is facing the front of the bench. Dress the the compost with a layer of fresh gravel or other decorative material,

if appropriate. Any staking, tying, or wiring required should be neat and must not detract from the appearance of the plant.

JUDGING NOTES Judges give preference to decorative rather than botanical value. Any flowers on a foliage plant are disregarded for their decorative value; foliage on a flowering plant is assessed as part of the overall condition and appearance. The highly coloured ornamental bracts (modified leaves) on plants such as *Euphorbia* (poinsettia) and *Bougainvillea* do not qualify as foliage.

JUDGES SCORE HIGHLY
Sturdy, shapely plants that are well furnished with healthy, unblemished foliage and displaying attributes true to the species and cultivar. Flowering plants with a good display of healthy, unblemished flowers or coloured bracts of good size, colour, and substance.

DEFECTS TO AVOID
Drawn, undernourished plants with unhealthy, deformed, undersized, scanty or diseased foliage of little ornamental value. Flowering plants with unopened and undersized flowers or bracts of poor substance and dull, ill-defined colours.

POINTS AVAILABLE

Condition	6
Quality and quantity of bloom, or decorative value of foliage	6
Cultivation	5
Difficulty of cultivation	3
TOTAL	20

Colours and markings of the foliage must be true to the cultivar

Plants should have lots of open flowers

Rex-cultorum Begonia

African Violet

The Floral Artist & Show Organiser

My name is Maureen Hinton. I've lived in the same village for four decades and have been involved with my local show for over 50 years, first as competitor, then member of the committee, chairman, and now a director of the show.

SIZE MATTERS

I started flower arranging at a very early age, inspired by my grandmother's cousin, who had a garden nursery (the business is still in the family). And I also pressed wildflowers and leaves, which we were still allowed to do in those days. Now I'm mainly inspired by nature itself – its colours and the changing seasons.

I love everything about competing. My favourite classes are the "miniature" and "petite" floral arrangements. Creating the best possible display is challenging but so rewarding, and therapeutic too. Sometimes it doesn't all go to plan. A few years ago, I was disqualified in a miniature class at a national show because the judges said that, although the flowers were perfect, since they had opened overnight my display had become too big! In those days a miniature had to fit into a four-inch cube. I've never lived this down, but I've certainly learnt from it.

FLORAL MECHANICS

The secret of a great flower arrangement in competition is the interpretation of the class title, working with colour and the space allowed, and not using too many accessories. You also need to get right what I'd call "the mechanics": conditioning flowers and foliage (trimming stems correctly so they don't wilt, etc.), and making sure you've read and understood the schedule

Minor Details

"Miniature" class arrangements must not exceed 10cm (4in) all round, even after the flowers have fully opened.

Fount of Knowledge
Maureen is always on hand in the horticulture tent to answer questions as competitors set up their exhibits.

"**Although the judges' decisions are always final, I will put my opinions forward if I feel strongly enough.**"

A Question of Scale
"Miniature" and "petite" designs should seem like large arrangements on a tiny scale, and finding flowers and foliage small enough can be a tricky job.

properly so you know what is expected. Above all, you have to enjoy the experience. Most judges are fair, and I really only ever have one criticism: some can be swayed too much by their favourite colour!

THE SHOW MUST GO ON

These days, I do still find time to enter my floral art into competition, but I'm also very busy organising the shows with my committee. There are lots of meetings, as you can imagine. We write schedules for each section, stage props, book judges, and ensure that competitors have all they need.

Over the years, there have been some very funny incidents. One involved the local lads making off with a competitor's iced cake after the show. The following year, she got her own back by icing a fake cake made out of plaster, which the lads, true to form, sneaked off with, only to find that they couldn't cut into it! She still won the class for iced cakes that year.

I believe that we have a duty to keep the village show alive, but it's a challenge. As well as the time and money involved, there are few young people interested in horticulture these days – technology has a lot to answer for. Having said that, my three grandchildren love coming to the shows, so maybe there's a future flower arranger amongst them.

Cacti & Succulents

There is glorious diversity in these spiky, fleshy house plants, from the truly tiny, through striking and architectural, to downright otherworldly. Cacti are types of succulents and not normally classed separately from other succulents.

JUDGES SCORE HIGHLY

Healthy, well-balanced plants with a full set of unbroken spines (if applicable); leaves with a fresh bloom and that are free from any defects and abnormal marks or lesions (except if they are close to soil level); good-sized specimens for the particular species, hybrid, or cultivar; pots of an appropriate size for the plant.

DEFECTS TO AVOID

Undersized, unhealthy plants with damaged or missing spines, distorted or scarred bodies or leaves, or a defective bloom; evidence of previous or current pest infestation; poor presentation; a specimen of flowering size that shows no evidence of flowering, if it is in competition with one that does.

POINTS AVAILABLE

Condition	6
Maturity (age in cultivation)	5
Freedom from pests and diseases	2
Difficulty of cultivation	3
Rarity in cultivation	1
Presentation	3
TOTAL	20

PREP Use a soft paintbrush to remove any dust on the plant. If you dress the surface of the soil with grit, inspect it for cleanliness and, if need be, remove it and sprinkle fresh grit. Take care transporting plants with delicate stems and flowers, and wear thick gardening gloves when handling specimens with sharp spines.

PRESENT Make sure the pot is clean and in proportion to the size of the plant. If the specimen you are showing has a best side, stage it so that this side is facing forwards on the bench.

JUDGING NOTES While the condition, maturity, and rarity of the plant are important, judges prefer plants in flower or showing evidence that they have flowered. A species, hybrid, or cultivar that is difficult to cultivate is preferred to one that is not. The judging criteria given here are applicable to smaller shows. For specialist shows, British Cactus and Succulent Society rules may apply (see www.bcss.org.uk for more details).

Leaves should be fleshy, firm, and healthy

Echeveria

Prickly Pursuits
If you're interested in cacti, look out for a stand run by the local growers' society; displays can be spectacular and members will be on hand to answer questions.

Plants should be dust-free

Rebutia

Pot should be in proportion to the size of plant

Rain in the Desert

People often incorrectly assume that desert-type cacti and succulents need little or no watering. In the growing season – from mid-spring onwards as temperatures rise – give plants a good soak once a week, but let excess water drain away. In winter, desert cacti need a dry rest and should only be given a little tepid rainwater if they show signs of shrinking.

Present with a fresh dressing of gravel

Spines should be intact and undamaged

Ferocactus

Judges expect to see flowers on plants that are of a flowering size

Overall shape should be graceful and balanced

Discoloration and marks to the skin are permissible near the surface of the compost

Echinocactus

Echeveria

Bonsai

Bonsai, meaning a "tree in a tray" in Japanese, are miniature trees grown in pots. A good bonsai tree should be as healthy as a tree grown in its natural environment. Trees with naturally small leaves are best for bonsai.

JUDGES SCORE HIGHLY

A neatly trimmed, healthy tree, looking as natural as possible, with all parts in proportion; a strong, shapely trunk, merging naturally with the soil and tapering upwards; well-proportioned head of branches, evenly spaced and set on the trunk with a natural apex. Weed-, disease-, and damage-free trees with healthy foliage colour and no obvious scars or marks of training.

DEFECTS TO AVOID

An unbalanced tree with growth uncharacteristic of the species; a weak, badly shaped trunk; branches that are badly spaced, cut, scarred, or crossed; snagged, cut, or dead fibrous roots; flowers, fruit, or foliage out of proportion to the size of the tree; any signs of damage.

POINTS AVAILABLE

The RHS does not recommend a particular points scoring system. Instead, judges are advised to award prizes on the basis of the overall appearance and health of the specimen, and the competency and creativity of the bonsai artist.

PREP Remove any dead or damaged leaves and branches, and trim any excess growth to reshape. Clean the bark carefully with a toothbrush, if needed. Rocks or wood used in the display will also need cleaning, and watermarks on the pot should be rubbed off. Remove any weeds and sprinkle fresh decorative gravel or bonsai soil over the surface of the growing medium. Remove any training wires that are ready to come off, and make sure any others are as discreet as possible (see box, below).

PRESENT The pot should be in proportion to the height and girth of the tree. Pots for coniferous trees are generally brown, occasionally grey, and unglazed; deciduous tree pots are glazed and a muted colour. Glazed pots can be given extra shine by wiping on some baby oil and removing any excess with a soft cloth. Avoid unnecessary additional ornaments or decoration. Position your bonsai on the show bench so that the angle from which it should be viewed faces the front.

JUDGING NOTES Neat and unobtrusive wiring on the trunk and branches is allowed, provided it doesn't cut into the bark. It is also permissible for surface roots to fan out from the base of the trunk and gradually disappear into the compost.

All Done with Wires

Bonsai artists carefully manipulate plants in order to produce their vision of an ideal tree in miniature. Pruning helps craft the tree's basic structure, while wiring shapes the tree into desired positions. By twisting wires around a trunk or branch, the wood is gently bent in a particular direction, until it "sets" in its new position and the wires can be removed.

Forest grouping formed from multiple slim trunks

Rock planting and semi-cascade shape creates a dramatic scene

Stage bonsai with the side from which it should be viewed facing the front of the show bench

Ezo Spruce

Surface roots are allowed, provided they disappear into the soil

Use brown pots for conifer trees

Japanese Larch

Tiny fruits are perfectly in proportion

Browned leaves remain throughout winter and are a desirable feature

Use a muted, glazed pot for a deciduous tree

Japanese White Beech

Rounded head of branches imitates the shape of the full-sized tree in nature

Dwarf Crab Apple

Orchids

TOP TIPS
Orchids grown as house plants need an environment that's as close as possible to their native rainforest in order to thrive. Keep them out of direct sunlight, and re-create humid conditions by misting their leaves and aerial roots several times a week.

Many of the 100,000 different orchid species and hybrids are well-suited to growing indoors. The orchid signifies beauty in the language of flowers and can make spectacular exhibits.

PREP Make sure the plant is well-watered and misted in the run-up to a show. Minimise sun exposure to prevent scorching. Remove any faded, damaged, or dead leaves. Dust the tops of the leaves and wipe gently with lemon juice to remove any water spotting. Any pseudobulbs (swollen, bulb-like stems) should also be clean and free of any dead matter.

PRESENT Make sure the pot is clean and in proportion to the plant. A neat, unobtrusive stake is allowed to help show off the flower spikes to their best advantage. Assess the angle from which the plant looks its best and stage with this side facing the front of the bench. Label the plant with its full Latin name, if known.

JUDGING NOTES Judges tend to look for plant spikes with an expected, or higher than expected, number of flowers for the genus. Tree-dwelling and rock-dwelling orchids may be shown attached to pieces of wood and rocks, instead of in pots. For more information and a complete set of rules on showing orchids, contact the British Orchid Council (BOC; www.british-orchid-council.info).

Flowers should be fully open and with bright, clear colours

Surface dressing on the compost is permissible

Alpine Plants

True alpines grow in high mountainous regions, but for show purposes, any plant that is small, compact, and hardy enough for growing in an alpine house or rock garden may be exhibited.

PREP Specimens grown in pots plunged in sand should be carefully extracted and the pot cleaned, although a light covering of mould growth on clay pots is not considered a fault. Remove faded, dead, or damaged flowers and foliage. Refresh the top-dressing of gravel, if required. Exhibitors are not advised to exhibit any specimens planted directly into a rock garden or raised bed, as the root disturbance and change of environment caused by digging up and transferring to containers is likely to damage the plants.

PRESENT Assess the angle from which the plant looks its best and stage with this side facing the front of the bench. Label the plant with its full Latin name, if known.

JUDGING NOTES Specimens need not be mountain natives and may be herbaceous perennials, annuals, or shrubs, but they must be of a size suitable for an alpine house (unheated, well-ventilated greenhouse) or rock garden, and hardy enough to survive an average winter in a frost-free alpine house. Classes for alpine collections of 3 or more plants are common, in addition to classes for single specimens. Provided they are of good quality, judges will prefer plants that are rare in cultivation and difficult to grow.

JUDGES SCORE HIGHLY

A plant rare in cultivation and growing true to its character in nature. Many perfect open blooms in a plant grown for its flowers. Closeness and firmness in a cushion plant. Colourful foliage in a plant grown for the colour of its leaves.

DEFECTS TO AVOID

A plant that is too large or not hardy enough (see Judging Notes). A plant that does not conform to its character in nature. A flowering plant with few flowers or flowers that are not open or are past their best. A cushion plant that is loose or patchy. A plant grown for its coloured foliage but lacking colour.

POINTS AVAILABLE

Suitability	2
Rarity in cultivation	2
In character	2
Cultivation	4
TOTAL	10

Cushion-type alpines should be close, tight form

Dionysia

Flowering plants should display plenty of open flowers

Jamesbrittenia

Flowers fully open and with bright, clear colours

Oxalis

The Gardening Dynasty

My name is Carol Pacifico, and I compete in the houseplant classes at my local show. My daughter Louisa and I are members of our local horticultural society.

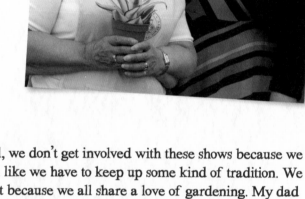

HOME GROWER

For the most part, I enter houseplant and flower classes. This year I entered a succulent into the show and won a first for it. I've been growing a few of those lately – my winning one is about two years old. I've also entered an orchid; got a third for that. I've got a lot of those at home. I think I have about 15 orchids at the moment, all around the house. And then I did a vase of garden flowers, and I got a second for that. Not too bad, really! Winning's not why I do it, though. I just like to take part.

I don't really do a lot of vegetable classes. My garden doesn't really lend itself to vegetable growing, you see. It all depends on the soil and the location of the garden. You've got to work with the land you've got; you can't force it. I do like to have a go at things, nevertheless. If they work, they work, and if they don't, they don't. I can normally do tomatoes and beans of some ilk, but I never got round to it this year!

A FAMILY AFFAIR

Gardening goes way back in our family. My father used to compete before I did – he used to exhibit at the Royal Horticultural Society shows. And, going even further back, Louisa and I have ancestors who were gardeners at Windsor Castle.

Still, we don't get involved with these shows because we feel like we have to keep up some kind of tradition. We do it because we all share a love of gardening. My dad was a gardener, I like gardening, Louisa likes gardening. It's something you pass on to your children: you get them

Succulent Success
Classes for cacti and succulents have become increasingly popular at local shows. Despite tough competition, Carol won first prize for her entry.

Child's Play
Novelty kids' classes, like Best Vegetable Animal, helped to spark an early interest in horticulture for Carol's daughter.

"I still have the cups in my cabinet from when Louisa won a prize for her flower and vegetable displays at just eight years old."

into it young, and it stays with them. You get them to grow a little something, or design a vegetable animal. Most shows have children's classes to nurture an interest in gardening in the next generation.

COMPETITIVE SPIRIT

Some people can get competitive when it comes to showing; they can take things quite seriously. If they don't get a first, they can get upset, wanting feedback.

It happens in all categories. I've seen bakers complain that the judge hasn't even cut into their cake, not realising that the cake they've baked hasn't been the right size. If the schedule calls for a nine-inch cake, the judges will go around with a ruler and measure every entry. And if they find a cake that's nine-and-a-half or nine-and-a-quarter, they don't even bother cutting it.

And, of course, when a class is popular, you can get a lot of disappointed exhibitors. One year, we had thirty-odd cakes of one type on the show bench – that's thirty-odd bakers all hoping for a first, when only one can take the trophy. You can put so much effort into your entry, it's not surprising that some people can be unhappy with the result. I'd like to think I'm not like that. I just like to get involved.

Bringing the Outdoors In
Since Carol's garden is not well suited to growing vegetables, she decided to turn her home into a houseplant paradise, including over a dozen orchids.

Primulas

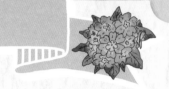

While you won't find a class for primulas listed in a schedule, classes for primroses and show auriculas are popular at spring shows and they are both types of (usually) pot-grown primulas.

JUDGES SCORE HIGHLY

Flowers of good substance with definite, clear colours. Primroses with compact growth and numerous flowers produced singly on long peduncles to form a symmetrical mass. Show auriculas with strong, long stems bearing trusses well above the foliage; with trusses carried on pedicels long enough to avoid overlapping of the pips; with a minimum 5 fully developed pips that are perfectly flat, round, smooth-edged, and with a smooth, circular, pure white paste.

DEFECTS TO AVOID

Plants with a loose habit of growth; foliage that is ill-balanced, limp, or showing pest or disease damage; unhealthy or blemished flowers of poor colour, with short, weak stalks or stems; too few flowers and any other deviation from the qualities described above.

POINTS AVAILABLE

SHOW AURICULAS

Foliage, stem and pedicels	7
Pips	2
Tube	2
Paste	3
Ground-colour	3
Edge	3
TOTAL	20

PRIMROSES

Condition	6
Floriferousness	5
Flowers	6
Colour	3
TOTAL	20

PREP Remove any faded foliage and flowers and give the pot a clean. Make certain you have correctly identified your primula. Primroses bear flowers on short single stalks, rather than in clusters on long stalks. Show auriculas are evergreen primulas with tight clusters of "salverform" flowers (trumpet-shaped with a flat face); each flower has a white circle in its centre, called the "paste".

PRESENT Assess the angle from which the plant looks its best and stage with this side facing the front.

JUDGING NOTES For both primroses and auriculas, judges are looking for plants in good condition, with well-balanced, healthy foliage, and undamaged flowers that are circular in outline and displaying clear colours. The judging criteria include some technical terms that are worth understanding: the "pip" is the individual flower in a cluster; the "truss" is a cluster of flowers; the "pedicel" is the stalk of an individual flower in a cluster; and the "peduncle" is the stalk of a single flower.

clear, distinct colouring to the flowers

Primrose bearing numerous single flowers on short stalks

Foliage is compact and well-balanced

clean pot

Tulips

Tulips provide bold splashes of colour in spring gardens. There are 15 botanical divisions, but most show schedules will only feature separate classes for single and double flower forms.

PREP Use a sharp knife to cut the selected stems. Some foliage must be left attached, but trim off any damaged outer leaves.

PRESENT Vases should be in proportion to the length of stem, size of flower, and number of blooms. Use packing material to place stems in an upright position; make sure multiple blooms are attractively arranged and well-spaced. No artificial support or wiring of blooms is allowed.

JUDGING NOTES All tulips, with the exception of those in the Single and Late Double and Parrot Groups, must have 6 petals and 6 filaments with anthers. Flowers deviating from this standard are not disqualified, but will only receive an award in the absence of acceptable exhibits. Any exhibit with flowers clearly diseased as a result of tulip-breaking virus will not be considered. Additional points are awarded for uniformity of size and form in classes calling for multiple blooms: up to 5 extra points are awarded in classes for 3 to 6 blooms, and up to 10 points for classes of 9 to 18 blooms.

JUDGES SCORE HIGHLY
Unblemished flowers in their most perfect phase of opening, of a good colour for the cultivar, of firm substance and smooth texture, and of the form typical of the division. Stems and foliage that are stiff and in good condition.

DEFECTS TO AVOID
Flowers immature or past their perfect phase, spotted, blistered, or otherwise blemished, of poor colour for the cultivar, thin, of rough texture, or of atypical form. Stems too weak to support the flowers and limp, badly blemished foliage.

POINTS AVAILABLE

Condition	4
Form	4
Colour	4
Size (for the cultivar)	2
Substance	3
Stems and attached foliage	3
TOTAL	20

Flower at the perfect phase of opening

Single, Cup-shaped Tulips

Leaves must be attached and should be fresh and unblemished

Stems must be stiff enough to support the flower

TOP TIPS

Some robust tulip cultivars flower year after year, but most are best regarded as annuals and should be lifted after flowering. Bulbs seldom flower well in the second year, but if replanted in autumn they may reach flowering size again in two years.

Irises

Depending on the show, these graceful beauties can be exhibited in pots or as cut flowers. A single class may be listed, or schedules may distinguish between irises grown from bulbs or rhizomes.

JUDGES SCORE HIGHLY

Sturdy stems with flower(s) fully opened; well-proportioned, fresh flowers of good colour; clean foliage.

DEFECTS TO AVOID

Stems that are weak; flowers that are damaged, unhealthy or fading; flowers not fully opened or few flowers open on stems with multiple blooms; blemished foliage.

POINTS AVAILABLE

Condition, including number of flowers open	5
Colour	5
Stem and foliage	5
Quality of flower	5
TOTAL	20

PREP Irises for showing in vases should be cut with a sharp knife, leaving as much stem length as possible. Harvest stems with the leaves still attached, or else cut some blemish-free foliage from the same plant and keep it with the flowers. Select uniform blooms if showing in classes for multiple stems of the same cultivar.

If showing potted irises, make sure the pot is clean and dress the surface of the compost with fresh gravel. Remove any leaves and flowers past their best.

PRESENT Vases for cut flowers should be in proportion to stem length, flower size, and number of blooms. Use packing material to hold stems upright. Ensure multiple blooms are well-spaced and attractively arranged. Position pots with the best side facing the front of the show bench.

JUDGING NOTES Judges may use technical terms to describe the parts of an iris flower, which are worth understanding. "Falls" are the three outer petals; these may have small caterpillar-like growths at the base, which are known as "beards", or a ridge of petal-like material called a "crest". "Standards" are the three inner petals of the flower, often smaller than the outer falls. More details about the many iris categories and guidelines for judging them can be obtained from the British Iris Society (BIS; www.british irissociety.org.uk).

TOP TIPS

For prize-winning irises, you need top-quality bulbs or rhizomes. They should be healthy and firm, with strong growing points and no soft or diseased areas. Smaller than average bulbs or rhizomes will not produce flowers in the first season.

Flower colour should be bright and clear

Show flowers that have opened completely

Aim to show stems with foliage still attached

Pansies

Pansies are a type of viola distinguished by the often large, dark patches on their petals. Garden and exhibition cultivars are usually judged separately, but are scored the same.

PREP Pansies are usually shown as single plants in pots. Remove any faded foliage and flowers, give the pot a clean, and give the plant a good soaking of water.

PRESENT Assess the angle from which the plant looks its best and stage with this side facing the front of the bench. Label the plant with its cultivar name, if known.

JUDGING NOTES Unlike exhibition cultivars, which are raised from cuttings, garden cultivars are generally raised from seed each year. This means a greater degree of variation will be seen in the appearance of the flowers of a particular garden cultivar, and judges will know to make allowances for this. It's worth understanding some of the technical terms used in the judging criteria: the "eye" is the centre of the flower; the "blotches" are the dark patches on the petals; "belting" refers to the margins of the petals outside of the blotches.

JUDGES SCORE HIGHLY
Large, clean, fresh flowers, broadly circular in outline, of harmonious colour, uniform in size and form, with symmetrical markings and smooth, velvety petals. Exhibition pansies: flowers with thick petals without serrations, lying evenly on each other and either flat or slightly reflexed; centre petals that meet above the eye and reach well up on the top petals; a deep, broad bottom petal that balances the others; belting of uniform width; blotches that are large, solid, rounded, and clearly defined; and an eye that is bright yellow, solid, circular and well-defined.

DEFECTS TO AVOID
Small, spent, blemished or damaged flowers, concave or lacking a circular outline, with poor-quality petals; a blotch lacking density or defined edges; an ill-defined eye that runs into the blotch; a V-shaped gap between the middle petals; the top edge of the bottom petal sloping down sharply. Exhibition pansies: serrated petals; belting too narrow or too broad, or of uneven width or ill-defined; blotches that are small, thin or ragged-edged; an eye that is dull or ill-defined.

POINTS AVAILABLE

Condition	3
Form and texture	5
Size	3
Colour	3
Belting	2
Blotch	2
Eye	2
TOTAL	20

Petals are smooth and velvety

Flower is broadly circular in outline

Garden Pansies

Markings show a good degree of symmetry

Carnations & Pinks

Both close cousins of the genus *Dianthus*, carnations are generally taller than pinks and have double flowers, while pinks include forms with single and semi-double flowers. What they share is a delicate beauty and fabulous scent.

JUDGES SCORE HIGHLY
Fresh, strongly scented, symmetrical flowers that are circular in outline, with firm petals, clear, bright colours, and well-defined markings.

DEFECTS TO AVOID
Flowers that are small or misshapen, with split calyces and with poor or no fragrance. Weak stems. Lack of uniformity in multiple blooms.

POINTS AVAILABLE

Condition	6
Form (of flower)	6
Size	3
Colour	3
Presentation	2
TOTAL	20

PREP Cut stems longer than needed, for trimming later. Cut foliage sprigs from the same plant and keep together with the stems. Remove wire supports and any bands on the calyces (flower cases).

PRESENT Insert stems and foliage into clear glass vases filled with packing material.

JUDGING NOTES Carnations are classed into two main groups for judging: "border" and "perpetual". Border carnations should have flowers with no hole or gap in the centre. Petals should be smooth-edged, but a slight indentation is permitted. Guard petals (outermost) should be large, broad, and smooth, and carried at right angles to the calyx, while the inner petals should lie regularly and smoothly over them. Centre petals may stand up and form a crown. Perpetual carnations should have large flowers with full centres. Guard petals should be flat, firm, and well-formed, although the edges may be smooth or regularly serrated. Pinks should have light and dainty flowers, with flat petals and smooth or regularly serrated edges. Guard petals should be broad and at right angles to the calyx. Single pinks must have five evenly shaped, overlapping petals. In double pinks, the inner petals should be evenly disposed, becoming smaller towards the centre.

Stems should carry subsidiary buds

Foliage should be "glaucous" (blue-green with a white bloom)

Flowers should face upwards or tilt at a slight angle

Double Pink (Laced)

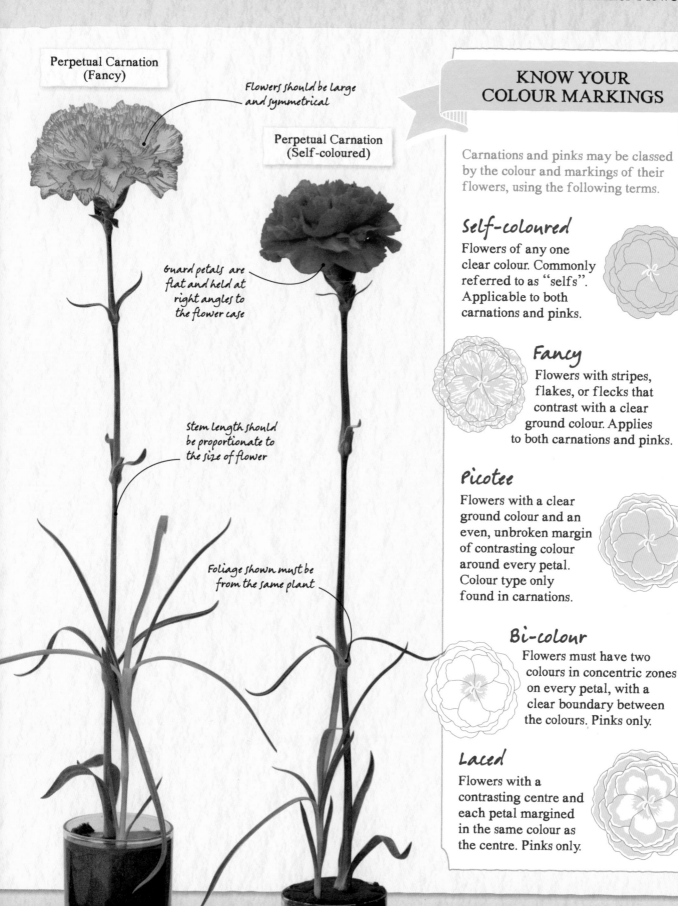

Perpetual Carnation
(Fancy)

Flowers should be large
and symmetrical

Perpetual Carnation
(Self-coloured)

Guard petals are
flat and held at
right angles to
the flower case

stem length should
be proportionate to
the size of flower

Foliage shown must be
from the same plant

KNOW YOUR COLOUR MARKINGS

Carnations and pinks may be classed
by the colour and markings of their
flowers, using the following terms.

Self-coloured
Flowers of any one
clear colour. Commonly
referred to as "selfs".
Applicable to both
carnations and pinks.

Fancy
Flowers with stripes,
flakes, or flecks that
contrast with a clear
ground colour. Applies
to both carnations and pinks.

Picotee
Flowers with a clear
ground colour and an
even, unbroken margin
of contrasting colour
around every petal.
Colour type only
found in carnations.

Bi-colour
Flowers must have two
colours in concentric zones
on every petal, with a
clear boundary between
the colours. Pinks only.

Laced
Flowers with a
contrasting centre and
each petal margined
in the same colour as
the centre. Pinks only.

Delphiniums

These midsummer stars of the cottage garden bear striking floral spires that always command attention. Schedules may include classes for single cultivars and multi-vase displays.

JUDGES SCORE HIGHLY

Healthy, upright spires that are long, tapering, or columnar in shape, with well-formed, well-coloured florets two-thirds open and evenly placed.

DEFECTS TO AVOID

Under-developed, malformed, or crooked spires; small, overcrowded, unevely placed or sparse florets, or any with faded or fallen petals.

POINTS AVAILABLE

Condition	5
Form and size of spike	5
Florets	5
Overall effect	5
TOTAL	20
Additional points in multi-vase classes:	
Uniformity	3
TOTAL	23

TOP TIPS

Delphiniums need to be planted in a sheltered position and given support as they grow, to stop the large spires from breaking in the wind. The ideal set-up is a frame of bamboo canes supporting several tiers of flower rings.

PREP Wait until most of the florets have opened before cutting delphinium spires. Cut some sprigs of foliage from the same plant and keep them together with the spires. Condition flowers by filling the hollow stems with water: plug the stem with cotton wool and tie a rubber band around the base to keep the plug in place and prevent the stem from splitting. Remove any dead leaves and florets.

PRESENT Display spires singly in tall, thin glass vases and insert clean foliage to conceal packing material; there must be no supports above the vase. Flowers must have at least 100mm (4in) of stem visible below the bottom florets.

JUDGING NOTES Good presentation and circular florets with neat and even "eye" (central) petals are preferred. There should be no signs of stripped florets or conspicuous seed pods. Extra points for uniformity are awarded in multi-vase classes.

At least two-thirds of the florets must be fully open

Florets should have a circular outline and neat "eyes"

Florets are evenly spaced along the length of the spike

Sweet Peas

With their delicate petals and delightful scent, sweet peas are a highlight of the summer garden. 'George Priestley', 'Jilly', 'Gwendoline', and 'Ethel Grace' are excellent exhibition cultivars.

PREP Cut flower spikes with secateurs, leaving a good length of stem for final trimming at the show. If exhibiting at a larger show, cut some sprigs of foliage from the same plant and keep them together with the spikes.

PRESENT Flowers at larger shows are normally presented in green "bikini" vases, with fresh, clean foliage from the same cultivar inserted to conceal the packing material. At smaller shows, sweet peas can be staged in mixed bunches, without foliage, in any attractive and suitably sized vase.

JUDGING NOTES Larger shows may feature classes for single cultivars, or mixed classes of 3 or more cultivars. Smaller shows usually ask only for a single bunch of mixed or single-cultivar blooms. Presentation is particularly important if competition is close.

JUDGES SCORE HIGHLY
Strong, straight stems in proportion to the large, well-spaced blooms; unblemished, fully open blooms of bright, clear colour and silken sheen.

DEFECTS TO AVOID
Weak or crooked stems; irregularly spaced flowers that are small for the cultivar, malformed, spotted, or with poor or running colour.

POINTS AVAILABLE

Freshness, cleanliness, and condition	7
Form, placement, and uniformity	6
Trueness of colour	4
Size of bloom in balance with stem	3
TOTAL	20

Mixed vases are usual at smaller shows

Blooms should be fresh and fully open

Glass vase shows off freshness of stems

Know Your Petal Parts

Sweet pea flowers are formed from three different petal types known as "standards", "wings", and "keel". Standards are the large petals at the back of the flower; the wings and keel form the front part of the flower, with the keel in the middle flanked by the two wing petals.

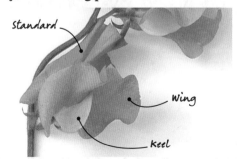

standard

wing

keel

Roses

Competition amongst top rose growers can be exacting – one British champion was relegated to second place for having a spider on the back of his bloom! The contest at your local show is likely to be less fierce.

JUDGES SCORE HIGHLY

Flowers at "perfect" or "full bloom" stage (see Know Your Rose Terms, opposite). Outer petals should regularly surround a well-formed centre typical of the cultivar. The blooms of a cluster rose should be gracefully arranged and so spaced as to permit their natural development.

DEFECTS TO AVOID

Immature, over-developed, or aged blooms; too many unopened buds; hips or stalks of spent blooms left showing. Untidy or unattractive presentation and evidence of petals having been removed or trimmed. Flowers or clusters crushed together or too widely spaced apart.

POINTS AVAILABLE

Form and size of bloom(s), form of cluster(s), substance	5
Freshness, brilliance, purity of colour	5
Stems and foliage	3
Presentation	7
TOTAL	20

PREP Use secateurs to cut roses, leaving a good length of stem with plenty of foliage. Remove any dirt or insects from blooms by gentle use of a soft brush.

PRESENT Arrange multiple stems so as not to crush blooms, create excessive gaps, or expose expanses of stem or foliage that detract from the flowers. Use of additional foliage will disqualify an exhibit. The heads of large-flowered roses may be held erect by a single wire.

JUDGING NOTES The following criteria are applicable to all roses. Petals must be firm, smooth, and a good texture, neither coarse nor flimsy, and blemish-free; their number should be typical for the cultivar. Blooms should be clean and sparkling, a good size for the cultivar, with no signs of tiredness or unnatural preservation. Colours should be glowing and bright, displaying the full depth of the true seasonal colour of the cultivar. Stems must be straight, and proportionate in thickness and length to the size of the bloom they support. Foliage should be adequate in quantity and size, as well as undamaged, fresh, clean, and of a colour and substance that is representative of the cultivar. Contact the Royal National Rose Society (RNRS; www.rnrs.org.uk) for information on different rose types and full judging criteria.

Arrange stems so that there are no excessive gaps in the foliage

Floribunda
Cluster-flowered

Miniature
Hybrid Tea

Bright, deep-
coloured petals

Miniature
Cluster-flowered

Stage roses
in a suitable
vase, such as
a bikini vase

Large-flowered
Hybrid Tea

Exhibition
roses should be
in either perfect or
full bloom stage

Floribunda
Cluster-flowered

KNOW YOUR ROSE TERMS

The rose exhibitor's lexicon contains
many different technical terms. Here
are just a few basic terms used to
describe the type, number, and
quality of blooms:

Single Bloom

A flower with fewer
than eight petals.

Semi-double Bloom

A flower with
between eight
and 20 petals.

Double Bloom

A flower with more
than 20 petals.

Perfect Stage

Blooms that are half
to three-quarters
open, with the
petals arranged
symmetrically within
a circular outline.

Full Bloom Stage

Blooms that are fully open,
with the petals arranged
symmetrically within a
circular outline. The
stamens, if visible,
should be fresh and
a good colour.

Begonias

Begonias are mainly tropical plants grown for summer bedding or as pot plants. Tuberous begonias, with their large, showy blooms, are the type most commonly exhibited at shows.

JUDGES SCORE HIGHLY

A well-balanced plant, with flowers of a proportionate size and number, and stout, erect stems; undamaged, large blooms, circular in outline, with petals culminating in one centre; petal colour of "picotee" cultivars (see p.143) must not run.

DEFECTS TO AVOID

Spindly plants with few flowers and too small for the cultivar; irregular blooms with a divided centre and long, narrow petals; pale, damaged, or spotted flowers and foliage.

POINTS AVAILABLE

Plant	5
Stems	3
Form of flower	6
Colour	3
Foliage	3
TOTAL	20

PREP Remove any damaged or dead flowers and leaves. Clean leaves by spraying with water, but keep the blooms dry. Unobtrusive, tidy supports are allowed.

PRESENT Position the plant so that its best view is facing the front of the bench. Make sure the pot is clean and undamaged, and in proportion to the plant.

JUDGING NOTES Competitors must submit 1 plant in its pot. Plants are only judged for appearance by the side on display and not for "all-round effect". The judging criteria given here apply to tuberous begonias; rhizomatous and rex begonias grown for their leaves can be entered into a class for foliage house plants (see p.127).

Petals should be broad and overlapping

Stems should be strong and upright

Flower heads can be supported by canes

Absolute Begonias

Start new tubers into growth six months before the show date, and old tubers five months before. Give plants a weekly high-nitrogen foliar feed (sprayed directly on leaves) and, once established, a high-potash fertiliser watered around the roots. Change from a nitrogen to a general fertiliser four months before the show.

Gladioli

Gladioli are bulbous plants prized for their tall spikes of funnel-shaped blooms. "Primulinus" gladioli have smaller and more loosely packed flowers, and are judged in a separate class.

PREP Select non-primulinus spikes one third in full flower, one third buds in colour, one third in green bud. Choose primulinus spikes with 14 to 20 flowers and buds.

PRESENT Show single spikes, either in tall, narrow vases or in their growing pots.

JUDGING NOTES Judges will prefer complete flower spikes, but up to two florets may be removed to improve appearance. The following merits and defects are specific to each gladiolus group: non-primulinus gladioli spikes should still carry the bottom bloom. Flowers should hide the stem and gradually narrow from base to top. Spikes should not appear to have too many blooms, but nor should the stem be visible between them. Primulinus gladioli spikes should be slender but strong, carrying 14 to 20 flowers and buds, facing forwards in a light, graceful stepladder arrangement; the upper petals should hood over the centre. Flowers should not look heavy and must not be so tightly positioned as to hide the stem.

JUDGES SCORE HIGHLY
Long, well-balanced, erect spikes with fresh, unblemished blooms and foliage; numerous regularly spaced open and opening flowers; flowers typical of the cultivar and of good form, texture, and colour, free from uneven and irregular markings.

DEFECTS TO AVOID
Short, bent, drooping, ill-balanced or crowded spikes; spikes with too few open flowers, too many empty bracts (bud cases), and blemished foliage. The removal of more than two blooms and more than one bract are major faults.

POINTS AVAILABLE

GLADIOLI,
NON-PRIMULINUS GROUP

Condition	6
Length and form of spike	6
Flower size, form and texture	4
Colour	4
TOTAL	20

GLADIOLI,
PRIMULINUS GROUP

Condition	4
Length and form of spike	6
Flower size, form and texture	6
Colour	4
TOTAL	20

One third of buds should ideally still be green

Non-primulinus Gladioli

Non-primulinus spikes must always still carry the bottom bloom

Foliage should be fresh and unblemished

Fuchsias

Fuchsias are summer-flowering shrubs with vivid, pendulous blooms. Shows often include several classes for the different growth habits and ways in which bushes can be trained.

JUDGES SCORE HIGHLY
Vigorous, balanced plants with many clean, fresh blooms and well-coloured foliage. Flowers should be fully open and complete with all floral parts, and should be at the point where pollen has just appeared or is about to appear. Standard fuchsias must be in proportion, with a length of clear stem within the specified range.

DEFECTS TO AVOID
Stunted, ill-balanced, sparsely covered plants, with insufficiently open blooms or blooms past their best; damaged or dirty foliage; untidy or obtrusive supports or ties; incorrect stem lengths; pest or disease damage.

POINTS AVAILABLE

Quality and quantity of bloom	8
Quality and quantity of foliage	5
Cultural quality	5
Presentation	2
TOTAL	20

PREP Pick off any dead flowers and leaves, and remove any nectar or pollen from the leaves by gently dabbing with a sponge.

PRESENT "Standard" forms (see Judging Notes) may be supported by a single stake. Neat and unobtrusive ties are permitted. Make sure the container is clean.

JUDGING NOTES Fuchsias can be shown as either bushes and shrubs, or in trained forms, most commonly a "standard". Bush fuchsias have single stems, which must not exceed 40mm (1½in). Shrub fuchsias are plants with more than one shoot and the shoots must come from below compost level. Standard fuchsias are specimens pruned and trained to form a bare upright stem supporting a head of foliage and flowers. Full standards have stems 760–1070mm (30–42in) long; half standards 460–760mm (18–30in); quarter standards 250–460mm (10–18in); and mini standards 150–250mm (6–10in).

Flowers should be abundant and fully open

Multiple stems indicate this is a bush form

Plants should have a balanced form

Pelargoniums

These South African natives are perfect for summer containers and will flower continuously if grown under cover. They are usually judged in two different classes according to type.

PREP Pelargoniums are judged for "all-round effect", so turn plants every few days to ensure a balanced display, especially in the run up to the show. Remove any faded foliage and flowers just before staging and give the container a final clean.

PRESENT Make sure the pot is clean and undamaged, and in proportion to the plant, and turn it so that the best side is at the front.

JUDGING NOTES Schedules usually feature two classes, one for ivy-leaved pelargoniums and another for zonal and regal types; both follow the same point system. Ivy-leaved pelargoniums are trailing plants with lobed leaves (separated into segments). Zonal pelargoniums have rounded leaves marked with a darker "zone". Regal pelargoniums are shrubby plants with serrated leaves and delicate, trumpet-shaped flowers.

JUDGES SCORE HIGHLY
Bright, clear, distinct flower colour and healthy, well-coloured foliage. Ivy-leaved types should be floriferous with a pleasing form, ample foliage, and well-developed trusses. Zonal and regal flowers should be large and round, with broad, overlapping petals.

DEFECTS TO AVOID
Misshapen plants with few leaves or flowers, or trusses that are not fully developed; coarse, yellowing, dull, or blemished leaves; weak stems. Zonal and regal trusses that are small, thin, or have too few full expanded flowers.

POINTS AVAILABLE

Condition	6
Trusses	8
Foliage	3
Presentation	3
TOTAL	20

Plant should be shapely and well-proportioned

Fully developed blooms with clear, bright colour

Healthy leaves with no signs of damage

TOP TIPS

To encourage flowering, move plants into a pot one size up every six weeks throughout the growing season, and feed with a high-potash fertiliser. Remove any dead flower heads and, every so often, pinch out shoot tips to encourage bushiness.

Tatton Park 2016

Pinks

CLASS   21

Second Prize

Tony Derrick

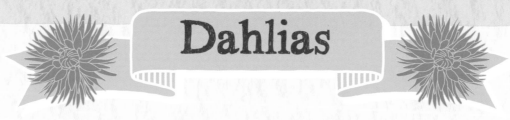

Dahlias

From midsummer to mid-autumn, dahlias tend to steal the show in the flower competition. Their fractal blooms are at the peak of perfection just before the petals start to fall, so timing is crucial when it comes to exhibiting.

JUDGES SCORE HIGHLY
Blooms with all florets (petals) intact, firm, and without blemish or defect. Colour(s) clear and well-defined, and evenly shaded or tipped throughout the bloom.

DEFECTS TO AVOID
Blooms that are malformed, face downwards, have limp, drooping florets or too many removed. Centres that are hard and green, undeveloped, or badly distorted.

POINTS AVAILABLE
Form and centre	5
Condition	10
Stem	3
Colour	2
TOTAL	20

PREP Read the full guidelines of the National Dahlia Society (NDS; www.dahlia-nds.co.uk) to be certain you have entered your dahlia into the correct class. There are a staggering 14 different dahlia groups and 9 further subdivisions by size, with only a small margin of error between giant and large, small and miniature. As well as errors in classification, displaying an incorrect number of blooms will also lead to disqualification, and it pays to remember that all flower buds, whether at embryo stage or showing colour, are treated as blooms and must be removed over the correct number.

PRESENT Blooms should face in the same direction, be clear of each other, and create a balanced effect; some leaves should remain on the stems. The names of all cultivars should be clearly stated. Blooms given artificial supports above the top level of the vase will be disqualified.

JUDGING NOTES Competitors should consult the standards for each class of dahlia set out by the NDS, which are numerous and detailed. The most we can hope to provide here are a few notable judging criteria for some of the most common classes. Single and

Florets must be compact and dense towards the centre

Thickness of stem should in proportion to the size of bloom

Foliage should be fresh, with no signs of disease or damage

Small Decorative Dahlias

Medium Decorative Dahlias

Colouring of the florets must be true to the cultivar

Planning Ahead
Cut stems longer than you need so that you can re-cut them for size when staging the display, and bring spare flowers in case any suffer damage on the way to the show.

Collerette dahlias should have 8 or more outer florets (individual parts of a flower head), which can overlap but must not assume double formation. The inner florets, or "collar", of Collerettes must be no less than one third of the length of the outer florets. Waterlily dahlias should have fully double blooms and the face of the bloom should appear circular in outline and regular in arrangement. Decorative, Cactus and Semi-cactus dahlias should have sufficient florets to prevent gaps in formation, but without overcrowding. Ball dahlias should be ball-shaped, but some flatness on the face of larger cultivars is acceptable. Outer florets must dress back to the stem to complete the ball shape of the bloom. Pompon dahlias must have perfectly globular blooms. The florets must be "involute" (turned in at the edges) for the whole of their length, and must dress back fully to the straight, firm stem.

Medium-large Semi-cactus Dahlias

Flower must be poised at an angle of no less than 45° to the stem

Ball Dahlias

Judges like blooms to be full but not overcrowded

Semi-cactus Dahlias

Cactus Dahlias

The Dahlia Exhibitor

My name's Jeffery Bennett and I'm a retired farmer. I grow Giant, Ball, and Semi-cactus dahlias for exhibition and I've won quite a few cups with my blooms.

A FLOWER THAT KEEPS ON GIVING

I didn't really do much gardening till retirement. When we were farming the land there was never time. Once we'd sold the farm, we started going to the big flower shows, at Chelsea and Hampton Court, and I found it was the dahlias that appealed to me. Dahlias are such big flowers and they will keep blooming from mid-July to the first frosts, which is a tremendous span.

I grow some dahlias under cover in a polytunnel and some out in the garden. The problem with growing dahlias outdoors is if you get a lot of rain at the wrong time the flower is spoilt by rain marks and can't be exhibited. But I do still like to see them outside. I dig my show tubers from the ground usually in November for storing over winter. At the end of February or early March I start to wake them up again with a bit of warmth and moisture. When I dig up the tubers, I put daffodils in at the same spot. As the daffodils come into flower the dahlias are just waking up, and when the dahlias are due to go in, the daffodils are ready to come out. That's my cycle. It doesn't do anything beneficial for the soil, I just like to see the garden busy.

> "I inspect the flowers and pick the ones I think will do best, then you're on a wing and a prayer hoping they won't go over too quick."

PERFECTION BEFORE THE FALL

I first entered my dahlias into small village shows and had some success, so I entered in the novice class at a larger horticultural show and won the cup for the best

Cut-and-come-again
Dahlias produce more flower stems after cutting, so you can show them again and again through the season.

Frostbitten, Twice Shy
After losing his entire stock to frost damage one year, Jeffery has learnt to give his tubers plenty of insulation during their winter storage.

> "Dahlias are absolutely beautiful flowers and they keep rolling on. Not like a peony where it's one bloom and then – bang! It's over."

two vases of "any of the above". I've won lots of cups since, but I'm always progressing; whatever standard you're at you always want to produce a little bit better.

Success is all about how well a bloom has developed by the day of judging. Dahlias are at their best when they are just about to go over and drop their petals, so a bit of luck is required to get it spot on. And I always take many more blooms to the show than I need as there are a lot of potholes in the roads round my way! I transport the flowers in crates now and each one is supported by a cane.

JUST A BIT OF FUN

We're a very friendly bunch of dahlia growers. We don't really see each other for the rest of the year, but on show day everyone's very social. We have a chat, pick up new ideas, and there's always a bit of leg-pulling. There are never any disagreements as you know yourself when someone's blooms are better than yours, and usually it's just down to luck on the day.

I'm happy competing at this more local level, at what I'd class as "the fun shows". The big boys' shows would really be taking it up a notch and I'm not sure I want to get that serious with my hobby.

Proven Winners
Jeffery first chose the dahlias he now enters in show by seeing which cultivars were winning prizes and assessing whether he could grow them better.

Chrysanthemums

JUDGES SCORE HIGHLY

Substantial, brightly coloured florets, fresh to the tips; clean, well-sized foliage. Intermediate blooms with globular outlines and reflexed blooms with graceful heads and full centres; both roughly equal in depth and breadth. Incurved blooms that are globular or nearly so, with full centres and broad florets that are closely and regularly, or loosely and irregularly, incurved. Spray blooms that are evenly spaced, and uniform in size, development, and colour.

DEFECTS TO AVOID

Non-spray blooms broader than deep, lacking good "shoulders" (shaping of the upper sides), or with depressed centres. Florets of poor substance, spotted, or of dull colour or drooping. Spray blooms that are faded or showing colour variation.

POINTS AVAILABLE

LARGE/MEDIUM EXHIBITION

Form	5
Size	6
Freshness	6
Colour	2
Staging and foliage	1
TOTAL	20

SPRAY

Form	3
Freshness	4
Colour	2
Size	1
Spray quality	6
Overall effect	3
Foliage	1
TOTAL	20

At specialist shows you may find classes for each of the 30 different types of chrysanthemums, but at smaller shows the most common are large or medium exhibition types and sprays.

PREP Select the most perfect blooms with a good amount of foliage. Cut with secateurs, leaving as much length to the stems as possible for trimming at the show. A cut chrysanthemum may struggle to absorb water. To assist water-uptake, you can make an upwards slit of about 75mm (3in) in the portion of stem that will sit under water. Remove any dead leaves and florets just before the show. Leaves can be cleaned by spraying with water and gently wiping, but keep the blooms dry.

PRESENT Display in a vase proportionate to the number and length of stems being shown. Use packing material, such as "oasis", to position the stems. If showing multiple blooms, make sure they are neatly arranged and spaced so that the judges can easily inspect them. The blooms of large and medium exhibition varieties can be supported by rings, and neatly tied canes may be used to support the stems of all types, provided they are unobtrusive and do not detract from the exhibit.

Downwards-curving lower florets

Flower head roughly equal in breadth and depth

The majority of florets curve upwards

Large Exhibition Intermediate

JUDGING NOTES Check you are entering your blooms into the correct class. Large and medium exhibition chrysanthemums will have had their flower heads "disbudded", which is where multiple flower buds are removed from the stem to leave only one central flower head. They may be further subdivided into classes for the form of the flower head, which can be "incurved", "reflexed", or "intermediate". Incurved flowers have florets (individual parts of the flower head) that open from the base and curve upwards. Reflexed flowers have florets that open from the crown (top) of the flower head, curving downwards and inwards. Intermediate flowers have florets that mainly curve upwards, but with some of the lower florets curving downwards. Spray chrysanthemums are those that have undergone limited disbudding or none at all, so that they produce several different blooms per stem. Natural sprays have been allowed to grow much as nature intended, without any disbudding, whereas exhibition sprays will have been partially disbudded, for example where the central bud is removed to give a more rounded outline to the spray. If you are uncertain which class your chrysanthemums should be entered into, contact the secretary of the National Chrysanthemum Society (NCS; www.nationalchrysanthemum society.co.uk) who will be able to send you full details of the many different chrysanthemum categories and guidance on the qualifying size of blooms.

Spray blooms must all be from a single main stem

Disbudded Spray

Retain the foliage, which should be small and fresh

Show stems with a clear majority of open blooms

Florets curve upwards

Natural Spray

Large Exhibition Incurved

Awarding the Prizes

Visitors arrive as the prize-winners are
revealed. Time to soak up the applause
and enjoy the rest of the day!

Exhibitors only discover who has won
once the show opens to visitors. For the
lucky few, there is usually a public
prize-giving ceremony, with awards
handed out by local community figures.

With the competition over for another year, winners and losers alike head to the tea tent for some well-earned refreshments and a chance to chat and swap notes.

Carrot cake
£1·40

Visitors mill around the show tables, examining the winning exhibits and making their own judgements. If they think they could do better, they may even be inspired to enter next year.

HOMECRAFT

BAKE & PRESERVE

Whether your talents lie in baking or preserves, this chapter has everything you need to whip up a winning domestic exhibit, including tried-and-tested recipes.

Tips & Tricks
HOMECRAFT CLASSES

Thanks to the continuing popularity of home baking and a resurgence of interest in preserving, the homecraft tent is guaranteed to see stiff competition from old-hands and novices alike. Most points are awarded for taste and texture, but errors in presentation will be marked down and risk disqualification.

Preserve Classes

PREPARATION Check if the size of jar is stipulated in the schedule before potting up the preserve. If applying labels before transit, wait until the preserve has completely cooled; labels must identify the main ingredients (usually the full recipe name is sufficient) and date made. If tying a decorative card tag to the jar, it is best to attach it at the show to avoid the risk of crushing in transit. If taking more than one unlabelled preserve of a similar appearance, ensure you have a way of telling them apart, such as a discreet sticker on the base.

TRANSIT Jars of preserves are relatively easy to transport. Pack them into a box or container that has been well lined with tea towels to ensure the heavy pots don't shift around or crack against each other during transit.

PRESENTATION Stage on the show bench with labels clearly visible from the front. Affix labels to the jar if you haven't already done so.

Jammy Mix Up

Some shows offer mixed preserve collections, in which exhibitors are required to enter a variety of preserves. As with classes for single preserves, each jar should be individually labelled and dated. Use matching jars for a neat presentation.

Baking Classes

PREPARATION For most classes, except for enriched fruit cakes (see p.176), bake as near to the show day as possible. Ensure all schedule specifications are followed, from using the correctly sized tin to following the steps of a set recipe. If possible, bake more than the specified quantity, so you have a range of bakes and spares to choose from on the day.

TRANSIT Store bakes in a cake tin or similar airtight container lined with kitchen towel. Do not overfill a container with small bakes, as this may damage them. To transport a large cake, place it on the inverted cake tin lid and then lower the drum over it; this avoids the risk of damage when placing it in the tin and makes it easier to lift out at the show. Remember to bring any spare bakes with you, in case anything gets damaged en route. If icing your bakes at the show, ensure you bring the required equipment, as well as the icing in a sealed container.

The Icing on the Cake

Decorating cakes in the tent avoids the risk of messing up icing during transportation, but it does pose its own challenges. You will need to bring with you more than enough of your preferred icing or buttercream, as well as any equipment required – there will almost certainly be no access to a kitchen at the show area. You will also need to allow yourself plenty of time, both to decorate the cake neatly and to give it time to dry before the judging begins. Consider too that it can get very warm in the tent.

PRESENTATION Arrange small bakes neatly on a clean plate, with a doily underneath, if preferred. Large cakes and tarts should be placed on a plate slightly larger than the circumference of the cake. Bread loaves may often be staged directly on to the show bench. Label all bakes with the name of the recipe along with any allergens it may contain, such as nuts or wheat.

Collection classes for bakes are a recent innovation and starting to catch on

Victoria Sandwich

This most British of cakes was a Victorian invention, so good they named it after the Queen Empress herself. Made with only a few simple ingredients, the Victoria is a classic test of the home baker's skill as there really is nowhere to hide.

JUDGES SCORE HIGHLY

Golden, evenly risen sponges with a delicate flavour, light texture, and visible layer of jam.

DEFECTS TO AVOID

Dry sponge with air pockets, a lumpy surface, damaged edges, and no caster sugar.

POINTS AVAILABLE

Appearance	4
Texture	4
Flavour	12
TOTAL	20

PRESENT Victoria sandwiches for showing should not be iced or filled with cream or buttercream. Do not be mean with the jam, but there should not be so much that it drips down the sides. Some schedules stipulate paper plates, otherwise choose a flat white plate slightly bigger than the cake.

PERFECT BAKE The top of the cake should be smooth, without bubbles, baked to a light golden colour, and decorated with caster sugar. The layers must be evenly risen and equal in depth, and the edges and sides must be smooth and undamaged. The sponge ought to have a light, open texture but without any large air bubbles, and above all must not be dry. A delicate flavour is desirable, with no overpowering taste.

JUDGING NOTES Schedules may provide a full recipe to follow or may specify only the number of eggs to be used; as a general rule, for every medium egg you will need 55–60g (2oz) each of caster sugar, butter or margarine (not low-fat), and self-raising flour.

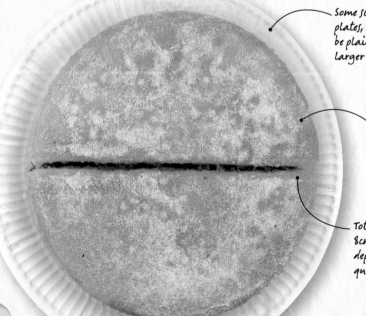

Some schedules call for paper plates, otherwise they should be plain, flat and 5cm larger than the cake

Decorate the top of the cake with a light, even dusting of caster sugar

Total height should be about 8cm and achieving this depends on the rise and the quantity of mix to tin size

TOP TIPS

For matching layers, weigh the tins before baking to check the cake mix is evenly divided. Turn sponges out the right way round to avoid marks from the cooling rack, and wait till they are cold before spreading the jam or it will seep into the cake.

Inspecting the Bake

Judges usually cut cakes in half to examine the texture of the sponge and distribution of jam, and then taste the sponge by removing a notch from the underside of the base layer.

VICTORIA SANDWICH RECIPE

Ingredients

175g (6oz) unsalted butter, softened, plus extra for greasing

175g (6oz) caster sugar, plus extra for dusting

3 medium eggs

1 tsp vanilla extract (optional)

175g (6oz) self-raising flour

1 tsp baking powder

115g (4oz) raspberry jam

Special Equipment

2 x 18cm (7in) round cake tins

1 Preheat the oven to 180°C (350°F). Grease the tins, and line the base and sides with baking parchment.

2 Whisk together the butter and sugar until light and fluffy. Add the eggs one at a time, beating well after each addition. Whisk in the vanilla, if using.

3 Sift together the flour and baking powder into the bowl. With a metal spoon, gently fold the flour into the mixture until completely combined.

4 Divide the mixture evenly between the tins and smooth the tops with a palette knife. Bake for 20–25 minutes until well risen. Test the sponges by inserting a metal skewer into their centres: they are cooked if the skewer comes out clean.

5 Leave the sponges to cool in the tins for 5 minutes before turning out. Remove the baking parchment and transfer the layers to a wire rack to cool completely.

6 Spread the jam over the top of the bottom layer, spreading right to the edges of the sponge. Carefully place the second sponge on to the first, lining up the edges, and lightly sprinkle the top of the cake with caster sugar.

Vegetable Cakes

If you find yourself with a glut of stump-rooted carrots or globe beetroot after growing a few for the horticultural show bench, why not grate your surplus crop and mix it up into a vegetable cake?

JUDGES SCORE HIGHLY
Moist, light sponge with a good balance of flavours, and of a colour appropriate to the ingredients used. Neat presentation and attractive decoration if filled and iced.

DEFECTS TO AVOID
Sponge that is underbaked and soggy or dry and overbaked, with a bland or unappealing taste.

POINTS AVAILABLE

Internal appearance	4
External appearance	4
Taste	12
TOTAL	20

PRESENT Unless the schedule states a preference, a vegetable cake may be either iced and filled or left plain. Cream cheese or buttercream icing is traditional, but be aware that dairy-based icing may melt (or even go off) left out in a warm tent all day. If no icing is used, the cake may be dusted with icing sugar to enhance the final presentation. Stage on a plate that is slightly larger than the circumference of the cake.

PERFECT BAKE Vegetable cakes should be moister than a standard sponge, but not overly soggy. Reduce excess liquid by draining the grated roots in a colander lined with kitchen towel, and pressing them gently before incorporating into the batter. Vegetable cakes are usually enhanced by spices or other flavourings, which should have a stronger taste in the finished cake than the vegetables themselves.

JUDGING NOTES Competitors must specify the vegetable used when submitting the entry. Judges will cut the cake in half, or remove a slice, to check the quality of the bake and taste the cake.

Use a fork to make concentric swirls in the icing for a simple decoration

Try decorating the top with an even scattering of chopped nuts

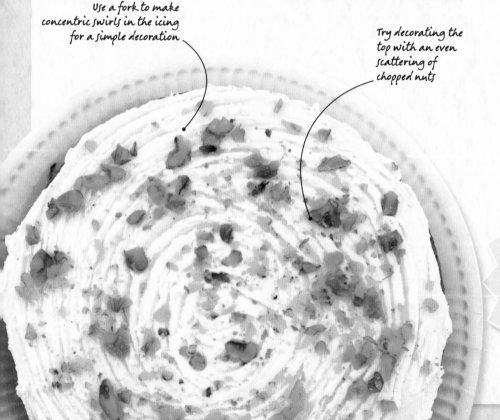

TOP TIPS
You can have great fun with this class. Carrot cake is the classic, but other veg work too: try beetroot in a chocolate cake; courgette with lemon; or even parsnip, which has a coconut-like taste that combines well with almond and vanilla.

Judges are likely to cut each cake in half to inspect the crumb

Feeding with Syrup

For a sponge that is evenly moist and flavoured throughout, pierce the surface all over with a skewer to aid the absorption of sugar syrup. Apply the syrup soon after the cake is out of the oven and while still warm.

Pierce the cake with a fine skewer to distribute the syrup

Drizzle Cakes

All drizzle cakes are finished by pouring a citrus-flavoured syrup over the top. Some rely solely on this syrup to create a sugar crust, while others call for an extra drizzle of water icing.

PRESENT Drizzle cakes are traditionally baked in a loaf tin. Some shows may allow other shapes, but check the schedule to avoid possible disqualification. Also check whether the class calls for a sugar crust-type cake or allows water icing. If using, take great care over the water icing to produce a pleasing pattern.

PERFECT BAKE Whatever drizzle cake you're making, the syrup must have fully penetrated the sponge to give an all-over moist crumb, and it must have a strong enough flavour to provide an obvious citrus tang, but without any bitterness. Lemon and orange are most common but any citrus fruit should work. If baking in a loaf tin, a crack along the top of the sponge is considered desirable.

JUDGING NOTES Judges will cut the cake in half, or remove a slice, to inspect the quality of the bake and taste the sponge.

JUDGES SCORE HIGHLY

A light, moist, even bake with good colour, and a strong citrus flavour; even distribution of syrup; crunchy sugar crust (if applicable); attractive water icing decoration (if applicable).

DEFECTS TO AVOID

Dry, underbaked, or overbaked sponge with a bland or unpleasant flavour. Poor penetration of drizzle. Lack of sugar crust or water icing, if specified in the schedule.

POINTS AVAILABLE

External appearance	5
Internal appearance	3
Flavour	12
TOTAL	20

The Cake Baker

My name's Sarahjane Luckham. I like to compete in the homecraft and handicraft sections at my local shows and I've won lots of classes and cups. I love anything food-related and baking is my passion.

FAMILY AFFAIR

I first got into baking when I was about five years old. My Nan, who was the best all-round cook, was my teacher and we used to bake together all the time. I'd class myself as a solitary baker now, but my Mum is a wonderful assistant if I ever need ingredients chopping. She's also great at clearing up after me – I tend to use every spoon I own! Me and my Dad share a couple of allotments and my first foray into competing was at our

allotment association's annual show. From there, I started competing in local village and country shows with my bakes. I also enter handicraft items alongside my Mum, while my Dad competes in the horticultural classes.

SECRETS OF SUCCESS

I think the secret to making the perfect vegetable cake is getting the mixture correct. Vegetables hold a lot of moisture and it's so easy to end up with a very soggy cake. I am perhaps proudest of my Beetroot Cake recipe. It suits any occasion, and everyone who tries it, loves it. It's also very versatile. For example, if you're

Cheese Cake
Sarahjane won the "Showstopper" class with this cake decorated in the style of a cheeseboard.

> "When I first won with a decorated cake, another baker had been winning the class for years and she was not best pleased a total newbie beat her!"

not keen on beetroot, you can reduce the quantity and put in more carrots, and if you are allergic to nuts, you can use sultanas in their place – it ticks all boxes.

LOOKING TO THE FUTURE

I think British village shows are fantastic, and I hope they continue for many years to come. I am really pleased that they have classes for children to enter. By starting young, they will hopefully get the bug and carry on competing. As for me, I would one day love to turn baking from being a hobby into my own small business. I like to read various recipes, take ideas from each one, and then merge it all to create something new. To publish a book of my own recipes is my dream.

> "My Nan taught me everything I ever needed to know about baking."

SARAHJANE'S BEETROOT CAKE RECIPE

Ingredients

250g (9oz) cooked beetroot, finely grated

100g (3½oz) carrots, finely grated

40g (1¼oz) walnuts, chopped, plus extra for decorating

250g (9oz) caster sugar

4 eggs

175ml (6fl oz) vegetable oil

250g (9oz) self-raising flour

2 tsp ground cinnamon

1 tsp ground ginger

1 tsp grated nutmeg

300g (10oz) icing sugar, sifted

150g (½oz) butter, softened

1-2 tsp vanilla extract

a splash of milk, if needed

Special Equipment

2 x 20cm (8in) round cake tins

1 Preheat the oven to 160°C (325°F). Grease the tins with a little of the oil and line the bases with baking parchment.

2 Put the beetroot in a sieve and press out the excess liquid until quite dry. Mix together with the carrots and walnuts.

3 Put the caster sugar, eggs, and remaining oil in bowl. Whisk for 5 minutes until light and fluffy. Fold in the beetroot mixture, then the flour and spices.

4 Spoon the mixture evenly into the tins and smooth level. Bake for 22-25 minutes or until a skewer comes out clean when inserted into the cakes.

5 Leave to cool in the tins for 5-10 minutes. Turn out on to a cooling rack, peel off the paper, and leave to cool completely.

6 To make the buttercream, put the icing sugar and softened butter in a bowl and whisk until combined. Add the vanilla extract and whisk again for 30 seconds. Add a splash of milk, if required, to loosen.

7 Spread half the buttercream on one sponge and place the other sponge on top. Ice the top of the cake with the rest of the buttercream, and decorate with a scattering of walnut pieces.

Spiced Beetroot Cake with Vanilla Buttercream

28/716

Fruit Cakes

The secret to fine fruit cake is a decent quantity of dried fruit that has been briefly boiled in water (or other liquids) and left to plump up overnight. An infusion of alcohol can add further richness.

JUDGES SCORE HIGHLY

A well-risen cake of good colour and moist texture, with an even distribution of fruits and a rounded spice flavour. If enriched with alcohol, the flavour should be consistent and well-developed.

DEFECTS TO AVOID

Underbaked or burnt cake with a dry texture or cracked surface; fruits that have sunk to the bottom of the cake, or that have burnt on the surface of the cake.

POINTS AVAILABLE

Internal appearance	4
External appearance	4
Taste	12
TOTAL	20

PRESENT Do not ice the cake unless the schedule specifically calls for it. The cake should have a gently domed, golden-brown surface once baked. If infusing your cake with alcohol and/or leaving it to mature, store it in an airtight container until show day. Present on a plain plate slightly larger than the cake.

PERFECT BAKE Fruit cake has a denser crumb than most, but should not be compacted. The top should be level or gently domed and the surface should be deep-brown in colour. The fruit should be evenly distributed. If using large fruits, such as dried apricots, chop them into smaller pieces so that they don't sink to the bottom.

JUDGING NOTES Check with the show organisers if it is unclear whether enriching the fruit cake with alcohol is allowed. Judges will cut the cake to taste, check the bake, and assess fruit distribution.

Watch out for fruits on the surface of the cake, as these can burn when baking

Judges will check for an even distribution of fruit by cutting the cake in half

This fruit cake gets its dark hue from the muscovado sugar in the mix

Boozy Fruit Cake

For a rich fruit cake, bake your entry up to three months ahead. Before baking, infuse the dried fruits in brandy or sherry overnight. Once baked, pierce several holes into the cake with a fine skewer, then "feed" it with 1 or 2 tablespoons of alcohol every few days for a month.

Judges want to see something that will make them smile!

Follow a theme, but be inventive

Bold, simple, well-executed designs will score highly

TOP TIP

Younger children might find it easier to work on rolled-out fondant cut to the shape of the bake. They can pipe their designs on to the flat surface of the fondant and leave them to dry, before fixing them to the bakes with a little buttercream.

Children's Bakes

You're never too young to enjoy baking, and classes for children's bakes are a great way to get kids involved in the kitchen. Decorating cupcakes or biscuits is a popular class.

PRESENT Classes often suggest a theme for children to follow, such as "farmyard" or "nature". It is better to choose a theme if none is offered, rather than create an assortment of unrelated designs. Fondant, buttercream, or water icing can be used, and can be either homemade or shop-bought.

PERFECT BAKE Children's classes usually offer set recipes that can easily be understood by children of primary school age.

The child should be able to follow the recipe independently, but a grown-up must be on hand to answer any questions and help with tasks such as chopping ingredients or using the oven.

JUDGING NOTES Classes typically call for 4-6 cupcakes or biscuits. Judges will take the child's age into account, as well as the execution and originality of design. Decorating classes do not usually judge the quality of bake.

JUDGES SCORE HIGHLY
Fun, creative designs of a uniform theme, which a child has clearly taken time and effort to produce.

DEFECTS TO AVOID
Unoriginal, messy, or unfinished designs of no apparent theme, or decoration that has clearly been done by an adult.

POINTS AVAILABLE

Neat presentation	3
Demonstration of creativity	10
Success of execution	7
TOTAL	20

Scones

These teatime stalwarts can seem deceptively simple to bake, but producing a uniform batch of show-worthy scones, with an even rise and fluffy texture, takes practice and a light touch.

JUDGES SCORE HIGHLY

Straight-sided, well-risen scones with a golden bake, fluffy texture and balanced flavour.

DEFECTS TO AVOID

Scones that have risen unevenly, are burnt or under-baked, are dry and tough or too crumbly, are bland or overpowering in their flavour.

POINTS AVAILABLE

External appearance	4
Internal appearance	4
Taste	12
TOTAL	20

PRESENT Check if the schedule specifies a size of cutter, otherwise 6cm (2½in) is usual. Brush the top of the scones with egg before baking for a good colour. Arrange the scones neatly on a plate.

PERFECT BAKE Make more scones than you need and cut one open to taste and check that you're happy with the bake. Scones must be well risen and not too lop-sided, evenly baked throughout with a light, fluffy texture. If flavouring scones the taste should be apparent, and multiple flavours should be in balance and complementary.

JUDGING NOTES Competition classes normally ask for 6 scones, but check the schedule to be sure. Sweet and savoury scones are judged in separate classes. Check if a sweet scone class specifies plain or with dried fruit, and allows other flavourings.

Scones should have good depth and relatively straight sides

Tops should be flat and golden in colour

Place the scones in a neat arrangement

TOP TIPS

Scones quickly go stale, so make your batch ideally less than a few hours before judging. The secret to light, tasty scones is to make the mixture a little wetter than you might think and only knead the dough very lightly before rolling it out.

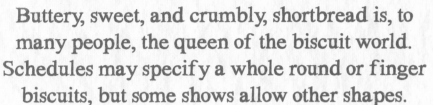

Shortbread

Buttery, sweet, and crumbly, shortbread is, to many people, the queen of the biscuit world. Schedules may specify a whole round or finger biscuits, but some shows allow other shapes.

PRESENT A shortbread round is usually marked into "petticoat tails" (wedges), but those made in a mould with a design can be left unmarked. A round should be left whole, but finger biscuits should be cut out. If pricking the surface with a fork or skewer, take care to apply the holes evenly. Lightly dust the surface with caster sugar.

PERFECT BAKE Shortbread should be baked to a medium golden colour. A round and fingers should be about 10mm (½in) thick and biscuits 6mm (¼in) thick. The texture should be crisp and crumbly, and it should melt in the mouth with a rich, buttery taste.

JUDGING NOTES Unless otherwise stated, "shortbread" means an entire round presented whole. Check the schedule to see if a specific number of fingers or other shaped biscuits is required.

JUDGES SCORE HIGHLY

An even bake to a golden colour; attractive presentation and uniform shapes and sizes; crisp and crumbly texture; rich, not overly sweet taste.

DEFECTS TO AVOID

Dark edges and a bitter taste from over-baking; texture that is too soft or too tough; messy presentation and uneven shapes and sizes to the portions or individual biscuits.

POINTS AVAILABLE

External appearance	4
Internal appearance	4
Taste	12
TOTAL	20

A Light Touch

Key to a crumbly shortbread is to avoid overworking the dough, as this causes glutens in the flour to bond too much, resulting in a tough texture. Mix in the flour very gently and stop as soon as it is incorporated, then press the dough into shape with your fingers rather than rolling it out.

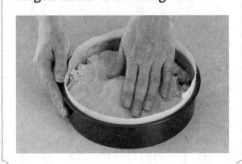

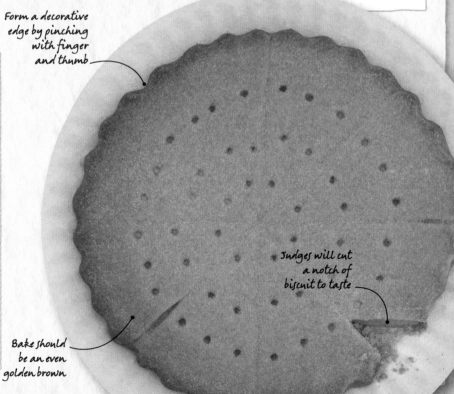

Form a decorative edge by pinching with finger and thumb

Judges will cut a notch of biscuit to taste

Bake should be an even golden brown

The Shortbread Baker

My name's Joy Chant and I enter my shortbread for competition (almost) every year. I run a small **B&B** and always leave some shortbread for my guests in the rooms. I've written out the recipe on many occasions.

BE-RO BEGINNINGS

I have always baked, ever since watching my own mother way back in the '60s as I was growing up. When I was at school and started Domestic Science lessons, I used to send off for all the books and leaflets that were on offer with various foods and baking ingredients. I still have some now and use my Art of Home Cooking (from Stork Margarine Cookery Service) and Be-Ro Home Recipes Book (cost 1/6d!).

I started competing at my local show in the 1980s. By entering in over 10 classes you would get a free ticket to the show, hence I started entering not just shortbread but cakes, jams, and other preserves as well, even sloe gin once or twice.

PLAIN AND SIMPLE

Shortbread is my mainstay. I am known to many as The Shortbread Lady, as I always turn up with some when there is a function, party, church sale, coffee morning, etc., going on.

I've no special secrets of success, just years of practice with disasters along the way. Shortbread must be made with real butter, nothing else gives such a good flavour. I have tried adding different ingredients over the years, but always return to my plain and simple, original recipe.

> **"Shortbread is my mainstay. I am known as The Shortbread Lady as I always turn up with some."**

ALL IN GOOD PART

I still enter my jams, chutneys, and lemon curd for competition, along with my shortbread, some years with better luck than others. There are always more clever people than I am.

It's a pleasure to win but I don't get too downhearted if I'm not a winner. Over the years you get to know the names that appear on winning entries. I've no ill-feeling, though, it's all taken in good part. One year a judge will look for certain characteristics then the next year go for something quite different. Each judge has their own thoughts. I may or may not agree with the judges' decisions, but I rarely change the way I bake and present my shortbread round. I have no more ambitions, really, to further my baking skills, not at my age!

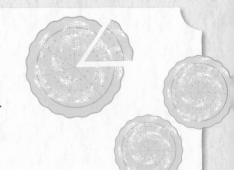

"The butter must be just the right texture, soft enough to make the dough but not too soft to handle and shape."

JOY'S SHORTBREAD RECIPE

Ingredients

250g (9oz) plain flour

85g (3oz) caster sugar,
  plus extra for sprinkling

175g (6oz) butter, softened,
  plus extra for greasing

1 If you're working by hand, mix the flour and sugar in a bowl and rub in the butter with your fingertips. I will own up! I simply put all the ingredients into my trusted food mixer and it does all the mixing.

2 Bring the mixture together to form a relatively crumbly dough. Briefly knead the dough on a lightly flowered surface until smooth.

3 If you are making "petticoat tails" – the traditional name for shortbread wedges – divide the dough in half and use your fingers to shape it into two 18cm (7in) circles, 2.5cm (1in) thick. Score each round into even wedges with a sharp knife and neatly prick the surface with a fork. To make shortbread fingers, roll the dough out into oblongs, prick with a fork, then straighten the edges and cut into fingers with a sharp knife.

4 Place the rounds or fingers on a greased baking sheet. Preheat the oven to 170-180°C (325-350°F) and bake for 20-25 minutes until pale golden. Get to know your oven as each one varies and will determine the cooking time.

5 I don't hurry to take the shortbread off the baking tray, but instead let it cool down slowly in the turned-off oven, which helps crisp it up. Sprinkle some caster sugar over the top, if you like, for extra sweetness and decoration.

Lemon Tart

A classic lemon tart is a thing of beauty: yellow as the sun, with a filling both sharp and sweet, held in a crisp pastry shell. Shows often have an individual class for lemon tarts and they always impress in a general sweet tart class.

JUDGES SCORE HIGHLY

A well-proportioned, neatly trimmed, attractively finished tart; rich, crisp, golden pastry that melts in the mouth; smooth, evenly set filling with a sweet-sharp lemon flavour and creamy finish.

DEFECTS TO AVOID

Pastry that's too thick, has shrunk too much and cracked, is burnt, tough, or hasn't baked all the way through; raw, rubbery, or split filling, with a weak or bitter taste. No leakage and no soggy bottoms!

POINTS AVAILABLE

Appearance	3
Pastry bake	4
Filling set	4
Flavour	9
TOTAL	20

PRESENT Bake with some pastry overhanging the tin and trim the excess with a sharp knife to achieve a neat edge. Prevent splits on the surface of the filling by leaving the tart to cool slowly in the oven with the door open. You can decorate the top with strands of lemon zest or icing sugar.

PERFECT BAKE The tart should not be too deep, but the pastry case must be well filled with the lemon mixture. The shortcrust pastry should be no thicker than 5mm (¼in) and evenly baked to a light golden hue, with a crisp, crumbly texture and rich, buttery taste. The filling should have a soft set right through to the middle, and hold its shape without being rubbery.

JUDGING NOTES Sweet pastry is more usual for a lemon tart, but they can be made with plain shortcrust; check the schedule to see if the pastry type is specified. The size of tart tin may also be stipulated. The lemon mixture is usually a baked custard, though some schedules permit filling with a homemade lemon curd; contact the organisers if the schedule is unclear.

pastry baked to an even golden brown

Filling evenly set with a strong colour

Judges may remove a small notch or larger slice to taste

Safe Lifting

Transferring the pastry to the tin can be nerve-wracking. Use the rolling pin to help you by carefully folding the pastry over the pin, then lifting the pin into place, centred over the tin, and unrolling the pastry. Do not stretch the raw pastry to fit the tin as this causes the pastry case to shrink as it bakes. Instead, roll the pastry wider than the tin and leave plenty of excess hanging over the edge, which can be trimmed later.

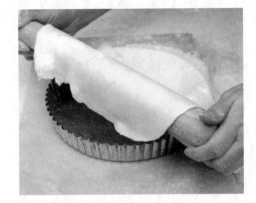

LEMON TART RECIPE

Ingredients

175g (6oz) plain flour, plus extra for dusting

85g (3oz) unsalted butter, chilled and diced

250g (8³/₄oz) caster sugar

6 eggs

Finely grated zest and juice of 4 lemons

250ml (9fl oz) double cream

Icing sugar and/or lemon zest, to decorate (optional)

Special Equipment

24cm (9¹/₂in) loose-bottomed tart tin

Baking beans

1 Mix together the flour and butter with your fingertips until it resembles fine crumbs. Stir in 50g (1³/₄oz) sugar, then add 1 beaten egg and draw it all together to form a pastry dough.

2 On a lightly floured surface, roll the pastry out thinly and to a circle large enough to line the tin with some overhang. Carefully transfer the pastry to the tin and press it firmly into place. Lightly prick the base with a fork and transfer to the fridge to chill for at least 30 minutes.

3 Whisk together the remaining eggs and sugar, then whisk in the lemon zest and juice, followed by the cream. Chill the filling mixture for 1 hour.

4 Preheat the oven to 190°C (375°F). Line the chilled pastry case with parchment, fill with baking beans, and bake blind for 10 minutes. Remove the paper and beans, and bake the case for 5 more minutes or until the pastry base is crisp.

5 Reduce the oven to 140°C (275°F). Place the tart tin on a baking tray. Carefully pour in the lemon filling and bake for 30 minutes or until just set.

6 Switch off the oven and open the door, but leave the tart to cool for 30 minutes or so in the oven, before turning it out of the tin on to a wire rack to cool completely. Transfer to a container for transporting to the show along with a little icing sugar and/or lemon zest to decorate, if desired.

Quiche

JUDGES SCORE HIGHLY
A generously filled, neatly finished quiche, with a crisp pastry case and a properly set, evenly cooked, flavoursome filling.

DEFECTS TO AVOID
Burnt, tough, overly thick pastry; raw or rubbery filling, with unevenly mixed ingredients; a bland overall flavour. Soggy pastry underneath.

POINTS AVAILABLE

Appearance	3
Pastry bake	4
Filling set	3
Flavour	10
TOTAL	20

What's the difference between quiche and tart? Some cooks believe it lies in the ratio of egg mix to other filling ingredients, but for village show purposes: a quiche is a tart is a quiche.

PRESENT Stage your entry on a relatively flat plate slightly bigger than the quiche. Ensure a neat top edge to the pastry by baking with an overhang and trimming away the excess with a sharp knife.

PERFECT BAKE The shortcrust pastry should be about 5mm (¼in) thick and evenly baked to a rich, golden brown with a crisp texture to both sides and base. The filling should have completely set and be full of flavour and well seasoned.

Ingredients in the filling should be evenly distributed, with the creamy egg mix dispersed throughout and binding everything together. Any cheese in the filling should have properly melted.

JUDGING NOTES Check if the schedule specifies a tin size; 20–30cm (7¾–12in) is usual. A quiche can often be enhanced with extra flavourings in the pastry, such as cheese, herbs, or spices, but check if permitted.

Top of the quiche is lightly caramelised

Pastry baked to an even golden colour

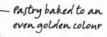

Quiche is well filled and ingredients are evenly distributed

TOP TIPS
To maintain even thickness of pastry, give it a 90° turn after each to-and-fro of the rolling pin, until you reach the size you need. Never turn the pastry over as this causes more flour to be incorporated, which can make the pastry dry and tough.

QUICHE LORRAINE RECIPE

Ingredients

150g (5¹/₂oz) plain flour, plus extra for dusting

75g (2¹/₂oz) unsalted butter, chilled and diced

1 egg yolk

200g (7oz) bacon lardons

1 onion, finely chopped

75g (2¹/₂oz) Gruyère cheese, grated

4 large eggs, lightly beaten

150ml (5fl oz) double cream

150ml (5fl oz) whole milk

Freshly ground black pepper

Special Equipment

23cm (9in) loose-bottomed tart tin, 4cm (1¹/₂in) deep

Baking beans

1 To make the pastry, mix together the flour and butter with your fingertips until it resembles fine crumbs. Add the egg yolk and 3-4 tablespoons of cold water to form a smooth, but not sticky, dough. Wrap in cling film and chill in the fridge for 30 minutes.

2 Preheat the oven to 190°C (375°F). On a lightly floured surface, roll the pastry out thinly and to a circle large enough to line the tin with some overhang. Carefully transfer the pastry to the tin and press it firmly into place. Lightly prick the base with a fork, then line with baking parchment

and fill with baking beans. Bake blind for 12 minutes. Remove the paper and beans, and bake for another 10 minutes or until lightly coloured and baked through.

3 Dry fry the bacon lardons on a fairly gentle heat for 3-5 minutes; they will release their fat. Add the onion and fry for a further 3-5 minutes until softened. Place the cooked pastry case on a baking tray. Spread the bacon and onion mix evenly over the case and sprinkle with the cheese.

4 Whisk together the eggs, cream, milk, and pepper to taste, then pour into the case. Bake in the oven for 25-30 minutes or until just set and golden brown. Remove from the oven and leave the quiche to cool briefly in its tin, before turning out on to a wire rack to cool completely.

A Crisp Base

There are several things you can do to prevent a dreaded soggy bottom: roll the pastry thinly; squeeze excess moisture from wet filling ingredients, like spinach; use a metal tin for good heat conduction. Most important, though, is to pre-bake the pastry case so that it is nice and crisp before the filling is added, a process called "blind baking". Weigh down the pastry with baking beans to prevent it puffing up, but then remove the beans and bake for a few more minutes to crisp the base.

Sausage Rolls

A savoury staple of picnics and parties, sausage rolls are traditionally made using flaky or puff pastry, though some prefer shortcrust for a more robust roll. Check if a pastry type is specified.

JUDGES SCORE HIGHLY

Rolls uniform in appearance and size, with correctly sealed pastry baked to an even golden brown colour (fully risen if layered), and a sufficient amount of well-flavoured sausagemeat.

DEFECTS TO AVOID

Rolls uneven in shape and size, with a patchy glaze or none at all; pastry that's too thick, greasy, soggy, or burnt; too much or too little filling with a bland or overpowering taste; no slashes and poor or incorrect sealing.

POINTS AVAILABLE

Internal appearance	4
External appearance	4
Taste	12
TOTAL	20

PRESENT Sausage rolls should have a neat seam where the edges of pastry meet. With layered pastry (flaky or puff), the join should lie along the side of the rolls so the layers can separate and rise. Seams of shortcrust pastry should lie beneath the sausagemeat, hidden from sight. Before baking, brush the pastry with an egg yolk glaze. Rolls should be 5–6cm (2–2¼in) long, unless otherwise stated, with three slashes on the top.

PERFECT BAKE The pastry should be light and crisp, with properly sealed seams and without any sogginess or greasiness. Rolls should have a balanced ratio of sausagemeat to pastry, and the filling should be flavoursome and moist without too much leakage.

JUDGING NOTES Check the schedule to see if a particular type of pastry is required. Rough puff, flaky, and shortcrust pastry must always be homemade, but show organisers will often allow full puff pastry to be shop-bought, given how labour-intensive it is to prepare. Also check if a minimum number of rolls must be submitted; 6 sausage rolls is usual.

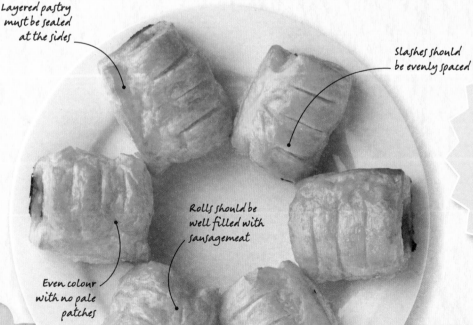

Layered pastry must be sealed at the sides

Slashes should be evenly spaced

Rolls should be well filled with sausagemeat

Even colour with no pale patches

TOP TIPS

Slashing sausage rolls before baking isn't just for decoration: the slashes allow steam to escape from the sausagemeat as the rolls bake, which helps prevent the pastry from becoming soggy. Over-filling the pastry is the main cause of sogginess.

Bread

White, wholemeal, seeded, sourdough: there is such variety in bread that shows often feature a basic loaf class and a freestyle class where bakers can more fully express their creativity.

PRESENT If a loaf tin is not being used, take sufficient care over shaping the bread so that it is neat, even, and rounded; a wet dough is more likely to lose its shape. Brush with a glaze of beaten egg before baking to add depth of colour to the crust.

PERFECT BAKE Bread should be well risen and fully baked, with a crisp, golden brown crust. The crumb should be even textured, springy, and soft. The dough should have had sufficient salt added to give a rounded flavour, and the crust should have a lightly caramelised taste.

JUDGING NOTES Check the schedule to ensure you are entering the right kind of bread. A loaf can be any shape unless a loaf tin is specified; a 1kg (2¼lb) tin is usual. Loaves made in breadmaking machines should only be entered in a class for machine-made bread. Loaves will be cut in half and rolls will be pulled apart to check the bake and for tasting.

JUDGES SCORE HIGHLY
Good shaping; a well-coloured crust with a pleasant crunch; an even crumb that's sufficiently light and airy; a rounded, fresh, yeasty taste.

DEFECTS TO AVOID
A rough, thick crust with large cracks or bubbles; crust that is overly hard or too soft, burnt or too pale; a bland flavour; a dense or uneven crumb and raw dough.

POINTS AVAILABLE

Appearance	4
Texture	6
Flavour	10
TOTAL	20

A Good Workout

For bread to rise the dough must be thoroughly kneaded. Elasticity in the dough comes from glutens in the flour, and kneading forces these glutens to form and stretches them out. Kneading also helps distribute ingredients and create a smooth texture. Proper kneading takes about 10 minutes hard work!

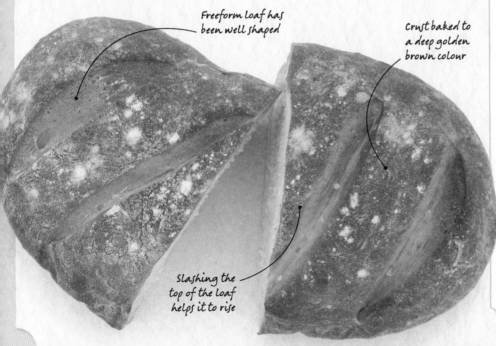

Freeform loaf has been well shaped

Crust baked to a deep golden brown colour

Slashing the top of the loaf helps it to rise

The Fancy Bread Baker

My name's Richard Brooks, I'm a smallholder and the Director of Human Resources at a university. I've entered my bread into competition for the past 10 years, both in the plain white loaf class and the "fancy bread" class.

KEEPING IT LOCAL

I live on a smallholding with my wife Angela, a Minister in the United Reformed Church, and our four dogs, two cats, and ponies. We also have sheep, pigs, chickens, and ducks, and our passion is being able to raise our own food. I particularly enjoy preserving foods: bottled tomatoes, homemade chilli sauce (I have about twenty different chillies growing in the greenhouse); bacon, ham, sausages, and pâté; cider and fruit cordials. It's not about self-sufficiency, it's about knowing where our food comes from and enjoying the process of creating it.

> "Village shows are about celebrating local community and celebrating people's talents, which we otherwise may never find out about."

Here in Southwest England there is a real focus on local food, with small shows springing up in villages all over. The last village we lived in had a show every two years and it gave people of all ages a reason to come together and enjoy some of the very simple, natural things of life: to create and to share.

BAKING EXPERIMENTS

I have been baking for about 20 years. I remember my mother used to make wholemeal loaves at a time when all you could buy was white sliced. Her loaves were very solid but wonderful eaten warm with honey.

I like to make a basic dough and then experiment by flavouring it with whatever I have around. Some home-grown, home-dried tomatoes from my polytunnel made a lovely flavouring chopped up with fresh herbs. I've also tried mixing seeds into wholemeal loaves. Sesame, pumpkin, sunflower, caraway: it's a great way to use up half-open bags left over from an enthusiastic trip to the health food shop.

The biggest lesson I've learnt is never to drop the salt straight on to the yeast. I used a breadmaker once to mix an initial dough, but I killed the yeast by putting the salt on top of it and ended up with paving stones!

Magic Dough
Richard loves seeing the results of the almost magical alchemy involved when dry flour and water transform into an elastic dough, given volume and life by the yeast.

TRIUMPH AND TAKING PART

I've probably won something every other year and it's great to win, but I'm not very competitive. I like to see how people have interpreted the class and it often gives me new ideas. The five-strand plait I entered one year resembled more of a lava flow than a delicate piece of bakery, but my bakes are mostly always edible! I usually enter some fruit or vegetables too, and often the men's baking class (though I'm fed up of trying to master a Swiss roll!). My greatest triumph was in a class called "A Cold Dessert": I made a raspberry mousse from fruit gathered in the garden and won first prize.

RICHARD'S TEAR'N'SHARE BREAD RECIPE

Ingredients

225g (8oz) strong white flour

225g (8oz) strong wholemeal flour

1 sachet dried yeast

1 tbsp dried milk powder

1 tsp salt

2 tsp sugar

300ml (10fl oz) lukewarm water

1 tbsp butter, melted

light oil, for greasing

1 tbsp smoked paprika

2 tbsp chopped fresh herbs, such as thyme, sage, and oregano

Special Equipment

30cm (12in) springform cake tin

1 In a large bowl, mix together the white and wholemeal flours with the yeast, milk powder, salt, and sugar until thoroughly combined. Make a well in the centre. Pour in the water and add the melted butter. Draw the flour into the liquid with a spoon and mix to form a soft dough.

2 Knead the dough on a floured surface for 5-10 minutes until smooth and elastic. Transfer to a lightly oiled bowl, cover, and leave to prove somewhere warm for about an hour, until doubled in size.

3 Knock back the risen dough and divide it into two equal pieces. Form one piece into a rough rectangle and sprinkle the paprika over the top. Fold in the sides of the dough and knead lightly to mix in the paprika, but not too much: you want to end up with a marbled effect.

4 Repeat the same process with the second piece of dough and the herbs, but this time knead thoroughly so that the herbs are well distributed.

5 Cut each half into eight equal pieces and roll each piece into a ball. Arrange the balls in concentric circles in the tin, leaving a little space between them and alternating the flavours. Leave to prove for another 30 minutes or so until the balls have joined up to form a single loaf.

6 Bake in an oven preheated to 190°C (375°C) for 25 minutes or until done. Turn out of the tin and leave to cool on a wire rack.

Chutneys & Sweet Pickles

These spiced, savoury tracklements both contain sugar and vinegar, but whereas good chutney relies on long, slow cooking, pickles preserve their crunch with a short cooking time and pre-salted or brined ingredients.

JUDGES SCORE HIGHLY

Jars properly sealed with the correct lid; thick, well-coloured chutneys with a complex, mellow aroma and flavour; tangy, balanced sweet pickles with the correct amount of bite to the ingredients.

DEFECTS TO AVOID

Jars sealed with plain metal lids; preserves that are over-spiced, bland or with a harsh vinegar flavour; chutneys that are too runny or too thick and overly caramelised; sweet pickles with ingredients that are too soft or too hard.

POINTS AVAILABLE

Jar and label	3
Appearance and set	3
Taste	14
TOTAL	20

PRESENT The vinegar content in these preserves will corrode metal, so seal jars with vinegar-proof lids with a plastic coating on the inside. Label jars with the date they were made and the main ingredients.

PERFECT PRESERVE Chutney should be thick enough to mound on the spoon, should be a rich dark colour, and have a balanced spicy-sweet flavour. The flavour of a sweet pickle should be tangier than a chutney but still balanced

and the vegetables should retain some bite. For both preserves, the ingredients may be chopped finely or left in larger chunks, but should be of a consistent size.

JUDGING NOTES Recipes often use the name "chutney" when it is properly a sweet pickle, which can lead to entries being disqualified. A simple rule of thumb is that if the cooking time is longer than 15 minutes, it is almost certainly a chutney.

Metal lids must be vinegar-proof

Jars should be well-filled

Ingredients should be chopped to a similar size

Chutneys should have a rich, dark colour

PICCALILLI RECIPE

Ingredients

1 large cauliflower, cut into small florets

2 large onions, finely sliced, or use small pickling onions

900g (2lb) mixed vegetables, such as courgettes, carrots, and green beans, cut into bite-sized pieces

60g (2oz) sea salt

2 tbsp plain flour

225g (8oz) granulated sugar

1 tbsp turmeric

60g (2oz) English mustard powder

900ml (1½ pints) ready-spiced pickling vinegar

Special Equipment

3 x medium glass jars with vinegar-proof lids

1 Put all the vegetables in a large, non-metallic bowl. Dissolve the salt in 1.2 litres (2 pints) of water and pour this brine over the vegetables. Put a plate on top of the vegetables to keep them submerged and leave for 24 hours.

2 The next day, drain the vegetables in a colander and rinse in cold water. Bring a large pan of water to the boil, add the vegetables, and blanch for about 2 minutes. Do not overcook them,

as they should be crunchy. Drain and refresh in iced water to halt the cooking process.

3 Put the flour, sugar, turmeric, and mustard powder in a bowl and mix in a little of the vinegar to form a paste. Put it in a large, stainless steel saucepan along with the remaining vinegar. Bring to the boil and stir constantly so no lumps appear. Reduce the heat and simmer for 15 minutes.

4 Place the clean jars in a low oven of 140°C (275°F) to sterilise them. Add the vegetables to the thickened sauce and stir well so that each piece is coated all over. Ladle into the warm jars, making sure there are no air gaps, and seal with the lids. Store in a cool, dark place and allow the flavours to mature for 2 months.

TOP TIP

Start well in advance of show day if you are entering a chutney or sweet pickle, as these preserves need time for the flavours to mature. Make at least two months in advance to allow the vinegar to mellow and the spices to soften.

The Judge of Preserves

I'm Vivien Lloyd and I'm an accredited judge of preserves. I also compete and my greatest success was winning "Best of the Best" at the Marmalade Awards; try my recipe on page 197.

LEARNING TO JUDGE

I first got into making preserves in my early thirties when I moved to a cottage with a very abundant garden in the village of Dodford in Worcestershire. A bumper crop of plums weighing 180kg (395lb) really kick-started my interest!

I entered my preserves at Dodford Village Show in 1993 and the judge at the show gave a talk to the local Women's Institute. After the talk she told me about a 12-week course in making preserves run by the National Federation of Women's Institutes (NFWI). Once I'd got my certificate I went on to train as an NFWI Preserves Judge. For a while the Women's Institute stopped running courses in preserves in an effort to get away from their "Jam and Jerusalem"

> "**As a judge I still like to compete to remind myself of what a competitor goes through to enter a show.**"

image. But a new wave of WI members keen to preserve and enter competitions has emerged and new courses have been introduced.

GIVING SOMETHING BACK

A good judge will have a clear, up-to-date knowledge of the full range of preserves, an ability to evaluate, and a desire to encourage exhibitors with helpful feedback. Judges should be impartial, suppress their own preferences, and avoid developing a personal resistance to a particular type of preserve.

I judge to give back the knowledge and skills others have given to me. I always learn something new when judging and sometimes I'm inspired to make a new preserve. I enjoy hearing about the success others have achieved in competitions, particularly if it is after reading my constructive comments. Quite often successful competitors go on to set up their own businesses selling preserves.

CHUTNEY OR PICKLE?

Judging has not changed over the years, but the skills and knowledge of many judges are different. Most of the generation of judges who taught me are sadly no

Label Recognition
Vivien will reluctantly mark an entry down if the main ingredients and the date it was made are not clear from the label.

"Shows often attract as many wasps as visitors, and judging with wasps in attendance requires a calm attitude, especially as I react badly to stings!"

longer with us. They (as I do) made the full range of preserves regularly and passed on their knowledge. It's worth noting that many village and country shows do not invite NFWI-approved judges, often because there's limited budget and a reluctance to pay for an accredited judge when a local resident will do it for free. As a result, judging standards vary and competitors often get mixed messages in their feedback or receive no feedback at all.

The biggest piece of advice I would give to entrants is to read the show schedule carefully and take note of the way each preserve should be presented, for example, if chutney needs to be shown with a new vinegar-resistant, twist-top lid. And don't get chutneys confused with pickles! I frequently find that jars labelled "Runner Bean Chutney" are in actual fact a type of pickle.

Sealed with a Disc
Always check the show schedule for instructions on how to present the preserves. If the schedule says curd must be sealed with a waxed disc and cellophane, any not sealed in this way will be disqualified.

Quiche Lorraine slice

Scones £1.00

Quiche lorraine
£1.20 slice

Scones £1.00

Victo...
£1...

Carrot cake
£1.40

Victoria Sponge
£1.00 slice

Marmalade

As everyone knows, marmalade is a favourite food of bears from Peru and marmalade sandwiches will sustain you over a long boat journey. Like jam, marmalade is a gelled fruit spread, but unlike jam it can only be made with citrus fruit.

JUDGES SCORE HIGHLY
Clean, correctly labelled jars with a good seal (unsealed jars may be disqualified); marmalade of a spreadable consistency and with a clear, strong aroma and flavour of the fruit used.

DEFECTS TO AVOID
Sticky jar with poor labelling; liquid on top; signs of scum in the mixture; a set that is too loose or too firm; unevenly cut peel; weak fruit flavour and/or a syrupy, caramelised taste.

POINTS AVAILABLE

Jar and label	3
Appearance and set	3
Taste	14
TOTAL	20

PRESENT Marmalade should be presented sealed in clear glass jars without trade names and with no stickiness on the jar or seal. Before potting up, remove any "scum" from the mixture and disperse air bubbles, then leave to cool slightly for 10 minutes to allow the peel to settle throughout the mixture.

PERFECT PRESERVE The preserve should have a wobbly, jelly-like consistency and an appealing, fresh colour. There should be no liquid or syrup on top. Citrus peel should be sliced, not minced, uniform in size and shape, and well distributed.

No scum should be present and there ought to be few, if any, bubbles. The fruit flavour should be fresh and obvious, and the peel tender. The taste should be a balance of bitter and sweet.

JUDGING NOTES Judges will open every jar and use a clean teaspoon to test the consistency and taste. They will also smell the preserve, hoping to find a strong aroma of the main ingredient.

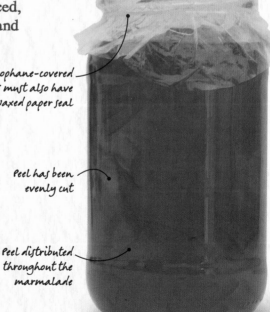

Cellophane-covered jars must also have a waxed paper seal

Peel has been evenly cut

Peel distributed throughout the marmalade

Why Seville Oranges?
The Seville orange is a rough-skinned type of bitter orange grown in Andalucia. Fruiting all too briefly from late December to mid-February, Sevilles have been highly prized on these Isles since Tudor times for their combination of bitter juice and excellent pectin levels, which together produce a quick set and fresh, tart flavour.

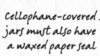

VIVIEN'S MARMALADE RECIPE

Ingredients

675g (1½lb) Seville oranges

1.75 litres (3 pints) of water

1 lemon

1.4kg (3lb) granulated cane sugar

Special Equipment

Large lidded preserving pan

Food processor

35cm (13in) square of muslin cloth

Sugar thermometer (optional)

5 x 450g (1lb) glass jars with new twist-top lids

Jam funnel (optional)

1 Juice the oranges and pour the juice with the water into the pan. Remove the inner membranes and pips from the shells of the oranges, but do not remove the pith.

2 Juice the lemon and add the juice to the pan. Place the orange membranes and pips, along with the remains of the lemon, into a food processor and chop finely.

3 Transfer this chopped mixture to the square of muslin cloth. Bring the sides of the cloth together and tie with string to form a bag. Add the bag to the pan. Shred the oranges and add the peel to the pan.

4 Bring the liquid in the pan to the boil, then turn down the heat, cover the pan and simmer very gently for 2–2½ hours, by

which time the peel will be very tender and the liquid should have reduced by a third.

5 Warm the sugar in an oven at 140°C (275°F) for 20 minutes. Remove the muslin bag from the pan mixture and use a large spoon to squeeze the liquid from the bag back into the pan. Add the sugar to the pan and stir until dissolved.

6 Bring the mixture to a rolling boil and test for a set after 7 minutes, using the flake test. Dip a large spoon into the pan and scoop out a spoonful. Lift the spoon above the pan and turn it horizontal. If the marmalade has reached the setting point of 104.5°C (220°F) it will drip, then hang on the side of the spoon. Alternatively, use a sugar thermometer to check when the setting temperature is reached.

7 Place clean jars in the low oven to sterilise them. Leave the marmalade to cool for 10 minutes, when a skin should have formed on the surface. Remove any scum with a large metal spoon. Gently stir the marmalade to distribute the peel.

8 Pour into the sterilised jars (a jam funnel will help) and cover with new twist-top lids. Leave jars upright and undisturbed to set.

Vivien Lloyd won "Best of the Best" with this recipe at the Marmalade Awards 2008. You can read Vivien's story on pages 192–3.

Jams

The very best jams capture the essential flavour of the fruit, and the way to achieve this is to avoid over-cooking. Boil jam for too long and you will end up with little more than a thick sugar spread.

PRESENT Fill jars just above the shoulder, sealing with screw-top lids or waxed discs and cellophane; 370ml (1lb) jars are traditional. Polish your jars to allow the colour of the jam to shine through. Labels should name the main ingredients and include the date it was made.

PERFECT PRESERVE The jam should be properly set but easily spreadable and have an obvious, fresh fruit flavour. Retaining a strong fruit flavour requires the shortest boiling time and if using fruit low in pectin, add extra pectin for a quicker set (see Perfect Preserve for jelly). Fruit pieces should be evenly distributed.

JUDGING NOTES Popular fruits, such as raspberries, may be given their own classes in a schedule. Unless particular fruits are listed, you are likely to see three main classes: "stoned fruit", such as plums, cherries, apricots, and peaches; "soft fruit", which includes all the berries, rhubarb, figs, and soft exotic fruit; and "hard fruit", such as apples, pears, medlars, and quince.

Jam should have a bright appearance

Lids can be decorated but must also have a proper seal

Testing for a Set
Using a sugar thermometer is the easiest way to test for a set: once the jam reaches 105°C (220°F) it should be ready to pot. Or try the "wrinkle test". Pre-chill a few saucers in the fridge. After boiling for 10 minutes, drop a teaspoon of jam on to a chilled plate. Leave for a minute and then gently push the mixture with your finger. The jam has set if the surface wrinkles and your finger leaves a trail.

Jellies

Like jams, jellies are sweet set preserves, but while jams are prepared with whole fruit, jellies use only the strained juice of simmered fruit for a more delicate, translucent preserve.

PRESENT Jellies are traditionally potted into 185ml (½lb) jars as they tend to break down and "weep" out juice if they are not used up quickly. Seal the jars with new screw-top lids or waxed discs and cellophane, and wipe clean. Once cool, stick on a label naming the main ingredients and date it was made.

PERFECT PRESERVE A jelly should have perfect clarity (see Top Tips), with a set that's more delicate than a jam, but not too soft, and a fresh fruit flavour. Pectin is the natural setting agent in fruit that makes the preserve jelly-like, but it is present in varying levels. If the fruit has low or medium pectin content, add extra liquid or powdered pectin at the same time as the sugar, to keep the boiling time short.

JUDGING NOTES As with jams, you may find three different classes for jellies categorised by fruit type, and/or a specific class for popular fruits like cranberries or apples (see Judging Notes for jam, opposite). Look out for sweet-savoury jelly classes, such as those made with herbs or chopped chillies.

JUDGES SCORE HIGHLY
Clean jars with a proper seal labelled with all necessary details; preserves with a wobbly, jelly-like texture and a clear, jewel-like consistency and colour; a distinctive, fresh flavour.

DEFECTS TO AVOID
Improper sealing and labelling; a set that is too runny; presence of "scum" (air bubbles from the vigorous boil); cloudy appearance; a weak or overpowering flavour.

POINTS AVAILABLE

Jar and label	3
Appearance and set	3
Taste	14
TOTAL	20

TOP TIPS

For the clearest jelly, do not squeeze the jelly bag to speed up the collection of juice, as this causes a cloudy appearance. And before potting the set jelly, skim off the "scum" (bubble layer) from the surface with a large, flattish spoon.

Labels can be printed or written out by hand

Jellies should be clear and translucent

Herbs and chilli pieces should be evenly distributed

...pple & Chilli Jelly
...e Sept 2015

The Jam Maker

My name is Debbie Pearce. I started making jam five years ago to use up a glut of home-grown fruit. I enjoyed it so much, I'm still doing it and now enter my jams into our local show.

UMBRELLAS AND HAIR CLIPS

It's very satisfying using produce you've grown yourself. I particularly love creating my own recipes and making jams you can't normally buy in the shops, like my Apricot and Passion Fruit Jam, which won first prize this year. My Mum has been my inspiration. She is a bit of a forager and as a child I remember rooting around the countryside with her. She'd go armed with an umbrella with a hooked handle and I'd have to untangle her when she got caught in brambles.

My toddler Orla enjoys making jams with me and I entered a show with our damson jam. She helped me pick the fruits from the garden and stone them. I told her if the jam won she could have the winnings. It did, and she used the prize money to buy some hair clips. We've also made pumpkin marmalade together this year using the bits left over from carving our pumpkin lanterns – seemed a shame to waste them!

KITCHEN DISTRACTIONS

The secret to perfect jam is quality ingredients and a decent preserving pan. But even so, disasters can still happen. Recently, my cousin Jayne gave me some lovely passion fruit from her garden "to do something with". I found a recipe on the net for mango and passion fruit jam and bought some fabulous ripe mangoes at the market. I let the fruits steep in sugar overnight. It all smelled great the next morning and I started making the jam, but was distracted by my baby

Plum Job
Debbie's damsons tend to ripen all at once and she needs to find creative ways to use up the glut.

DEBBIE'S APRICOT & PASSION FRUIT JAM RECIPE

Ingredients

1kg (2¼lb) apricots

750g (1lb 10oz) granulated sugar

Pulp from 10 passion fruits

Juice of 1 lemon

Special Equipment

Preserving pan or large, heavy-based saucepan

Discs of waxed paper

4 x 450g (1lb) glass jars with new metal lids

1 Stone and quarter the apricots, then mix them together with the sugar in a large bowl. Cover and leave to steep overnight. Place a few saucers in the fridge to chill; this is for testing if the jam has set.

2 The next day, pour the fruit mixture into a preserving pan, together with the passion fruit pulp and lemon juice.

3 Stir on a low heat until the sugar has dissolved, then turn up the heat and boil vigorously until setting point is reached; about 15–20 minutes.

4 To test for a set, remove the pan from the heat and put a little of the jam on a chilled saucer. Leave for a minute, then see if it forms a skin and wrinkles when you push it with your finger. If not, continue boiling and test again every few minutes.

5 Heat the jars in a 140°C (275°F) oven to sterilise them. Ladle in the jam, then cover with waxed paper discs and seal with metal lids.

son Roman. The sugar ended up caramelising and left the jam a horrible brown colour and looking more like chutney. It certainly wasn't going to win any prizes, but it still tasted good!

THE ONE TO BEAT

For me, taking part in these competitions is just a bit of fun, and I wouldn't advise anyone to take it too seriously. That said, I always think Mrs Lemon is the one to beat in the jam-making competition at our local show, and I do feel a sense of achievement if I get placed above her in a category.

Preserving Pumpkins
Rather than wasting her Hallowe'en lanterns, Debbie was inspired to turn the discarded flesh into pumpkin marmalade.

"As long as my trees and fruit bushes keep producing, I will carry on making jam!"

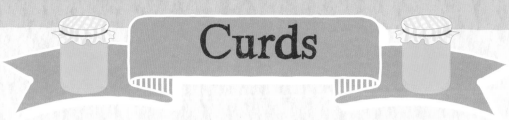

Curds

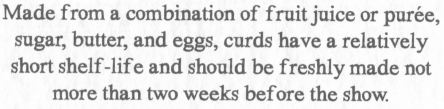

Made from a combination of fruit juice or purée, sugar, butter, and eggs, curds have a relatively short shelf-life and should be freshly made not more than two weeks before the show.

JUDGES SCORE HIGHLY
Correctly sealed, clean, properly labelled jars; curds with a bright colour, good set, smooth texture, and rich, tangy flavour.

DEFECTS TO AVOID
Unsealed jars; curd that's too runny, too thick, or split; bland flavour or bitter aftertaste.

POINTS AVAILABLE

Jar and label	3
Appearance and set	3
Taste	14
TOTAL	20

PRESENT Curds are traditionally potted into 185ml (½lb) jars (check the schedule to see if a larger size is stipulated), which should be sterilised by warming them in an oven set to 140°C (275°F). The curd should then be sealed with waxed discs and, once cooled, covered with cellophane and labelled with the main ingredients and date made. Store jars in the refrigerator until the day of the show.

PERFECT PRESERVE The fruit curd should be properly set to a spreadable consistency, though it shouldn't be stiff. It should have a bright colour, fresh aroma, and the taste should be rich and creamy with an obvious fruit flavour that is tangy and sharp, but without any bitterness.

JUDGING NOTES Curds are most commonly made with the juice and rind of citrus fruits, but they can be flavoured with many other fruits and you should check the schedule to see if a particular fruit is specified. Lemon curd is the classic curd and shows will often give it a class all to itself.

Jar should be sealed with waxed discs and cellophane

curd has a bright, pleasing colour

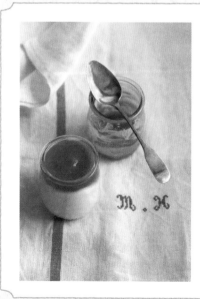

Beyond Lemon Curd
Curds are not just for lemons. Most other citrus fruits work well, such as pink grapefruit or orange with lime. Lots of non-citrus fruits are suitable for curd-ing too, provided they have a sharp enough flavour to cut through the richness. Try gooseberries, raspberries, currants, or rhubarb with vanilla. Nor is curd only good for spreading on scones: top plain yoghurt with a layer of curd or drizzle over meringues for a quick, indulgent dessert.

Fruit Liqueurs

Fruit liqueurs are made by sweetening spirit alcohol and infusing it with the flavour of fresh fruit. Judges will expect a strong but rounded taste of fruit with no harshness from the alcohol.

PRESENT Liqueurs must be presented in small, sterilised, stoppered bottles, which should be labelled with the ingredients and the date bottled. Make sure the bottle is full! Unless otherwise stated, place a small liqueur-style glass alongside the bottle.

PERFECT LIQUEUR The liqueur should be sweet, with an obvious fruit flavour, and a clear, bright appearance. Any additional flavouring from herbs and spices should complement the taste of the fruit; for example, juniper works well with damsons and vanilla in peach brandy. Use vodka for a bland base alcohol or gin when you want some aromatics. Don't skimp on the sugar or the liqueur won't be sweet enough for the fruit flavour to come through.

JUDGING NOTES Liqueurs must be made with spirit alcohol, which is defined as a distilled, unsweetened alcohol of at least 20% ABV, such as vodka, gin, or brandy. Sugar must be fully dissolved and there should be no fruit or detritus in the bottle.

JUDGES SCORE HIGHLY
Clean, properly labelled bottles; liqueurs with a bright colour, balanced flavour with fruit to the fore, sufficient sweetness, and a mellow finish.

DEFECTS TO AVOID
Improper labelling; bland or unbalanced flavour; harsh alcohol taste; a cloudy appearance.

POINTS AVAILABLE

Jar and label	3
Appearance	3
Taste	14
TOTAL	20

Worth Waiting For

Allow at least six weeks, but preferably three months, for the fruit to steep before straining the mixture overnight through a muslin bag; do not squeeze the bag as this will turn the liqueur cloudy. The longer you can leave the finished liqueur to mellow, the better. Ideally you should make it a year ahead of the show, but a longer period of maturation will always be beneficial – assuming you can wait that long!

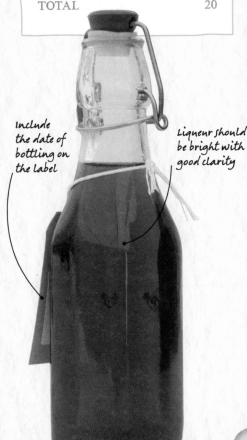

Include the date of bottling on the label

Liqueur should be bright with good clarity

The Show Organiser & Curd Maker

Judgement Time
After the buzz of competitors dropping off entries comes the quiet time when judges inspect, taste, and award prizes.

My name's Sue Fear, I work as a Teaching Support Assistant at an infant school and I run the Home and Handicraft section of my local show. I usually enter one or two classes as well, and this year I won top prize with my lemon curd.

RUNNING THE SHOW

I have been chair of the Home and Handicraft section for eight years and have been on the committee for 20 years. As a teenager I used to help set up and clear up the marquee. My mum and grandmother were also committee members. They were both farmers and my brother still runs the farm where we grew up – farmers are good organisers and we're happy mucking in.

This year I tried something new and got each helper to contribute a dish so we could put on a lunch for all the judges, who really do work hard. Our judges are always very thorough and fair and encouraging; many of them have been trained by the Women's Institute. The lunch seemed to be a success and brought everyone together. I think it turned the judging into an enjoyable day out.

BUZZ OF COMPETITION

I love seeing all the amazing exhibits coming in on the Friday morning, ready for the judging that afternoon. And I love the excitement on show day of those who've won classes and trophies.

I've had some success at the shows in the past. Ten years ago I won the trophy for most points in the homecraft section for baking and preserves. I don't

enter as much now as it would be very embarrassing to win a trophy while being chair of the section. But I do miss the buzz of competing.

PATIENT STIRRING

The lemon curd recipe came from my great aunt and I have fond memories of seeing her cooking it when I was a child. I make it for special occasions or to give as a gift. We eat it on bread or crumpets, and I use it as a filling for cakes. I have had problems in the past with the curd being too thin. But you can't hurry the process of setting the curd and you must never stop stirring or the mixture will overheat and separate.

> "We had 39 lemon drizzle cakes one year and the poor judge felt sick after tasting them all!"

SUE'S LEMON CURD RECIPE

Ingredients

3 unwaxed lemons

100g (3½oz) unsalted butter

3 eggs, plus 1 extra egg yolk

225g (8oz) granulated sugar

Special Equipment

2 x small glass jars

waxed paper discs, cellophane, and elastic bands

1 Finely grate the rind of the lemons, avoiding as much of the white pith as you can since this can give the curd a bitter flavour. Squeeze the juice from the lemons and strain it through a sieve to get rid of any bits, for a smooth curd.

2 Chop the butter into smallish pieces and melt it in a heatproof bowl set over a saucepan of gently simmering water. Make sure the water doesn't touch the bottom of the bowl as this can cause the curd to cook too quickly and it may separate.

3 Lightly beat together the eggs and extra egg yolk. Add the beaten eggs, sugar, lemon rind, and juice to the melted butter, and stir to combine.

4 Cook the mixture very gently without boiling, stirring all the time, until it thickens to a spreadable consistency. The curd is ready when it coats the back of a spoon and you can trace a path through it with your finger.

5 Place the clean glass jars in a cool oven at 140°C (275°F) for 15 minutes to sterilise. Pour the curd into the warm jars, seal the surface with a disc of waxed paper, and cover with cellophane held in place with an elastic band. Store the curd in a cool place (it will keep for a month) and keep it in the fridge once opened.

NUTS & BOLTS

For anyone interested in hosting a show in their own village or community, the following pages cover all the basics, from arranging the event to writing the rules.

Organising a Show

Most village shows are organised by local horticultural and flower societies. With a little time and planning, societies and groups of any size can put on a show of their own.

APPOINT THE SHOWRUNNERS

There are a number of roles involved in organising a show. The show secretary is usually responsible for the general organisation of the show. They are often responsible for appointing judges, arranging publicity, and deciding the show's rules as stated in the schedule (see pp.212–15). They usually appoint an official to receive entries and entry fees from exhibitors, as well as a group of stewards to help things run smoothly on the day of the show (see Stewarding Success, opposite).

PLAN THE BASICS

Visit other shows in nearby towns and villages to gather ideas and information. Before setting a date, consider any potential clashes with other events in the local area. Choose a suitable venue, bearing in mind that it should have running water, plenty of light, and be the right size for the anticipated number of exhibits. This is also the time to source any staging materials, such as tables, vases, and plates.

KNOW YOUR MARKET

Many different types of people exhibit at local shows. Talk to local groups such as gardening clubs, schools, and social clubs to gauge the level of expertise and to establish which categories and classes will be likely to attract any interest, and therefore should be included. Consider also

Finding the Judges

A good judge is impartial, has common sense, is familiar with the kinds and cultivars of the classes to be considered, and has knowledge of the skill required to grow and stage them. Local plant societies, county judges' guilds, and the RHS (www.rhs.org.uk/affiliatedsocieties) can help you locate horticultural judges, while the Women's Institute can be a good resource for domestic judges. Successful exhibitors often make good judges, too.

whether children's classes or novelty classes, such as giant vegetable competitions or men-only baking classes, may be appropriate.

CONSIDER THE PRIZES

First, second, and third prizes for each class are normally awarded, with a modest cash prize for each. "Highly Commended" or "Commended" awards can also be made. "Best in Section" and "Best in Show" awards are also often included, as are other special awards.

PUBLICISE THE SHOW

Work out how best to get the word out among your community. Consider putting out a press release for local media, or inform other local horticultural societies and gardening clubs. You could also think about putting up posters around the local area, or creating a website or social media profile for the event.

ORGANISE INSURANCE

Organisers of village shows are liable for injury suffered by anyone visiting or participating in the show, including members of the public, guests, exhibitors and voluntary staff. If your organisation does not have public liability insurance cover then the liability lies with the officers of that organisation. It is therefore essential that adequate cover be provided. The owners of the venue in which the show is held may sometimes provide cover; if not, the organisers must take out a special policy to cover the event.

Stewarding Success

Stewards are an important feature of any show: they act as the link between the judges and exhibitors, and are there to ensure that judging runs smoothly. Stewards can also assist new exhibitors, ensure entries are staged in the correct areas, check that all entries have been judged, and help with any other issue that may arise in the tent.

ESTABLISH THE RULES

The rules of the show must be made clear to all judges and exhibitors. The RHS's Horticultural Show Handbook offers suitable rules for most shows. All rules and judging information should be clearly stated in the show schedule to avoid any possible misinterpretation.

RUNNING THE DAY

Make sure that enough helpers and stewards have been appointed to help the day go smoothly. Ensure the doors open promptly at the published time for staging, and clearly explain to exhibitors how long they have to stage their exhibits. Ensure prize money is paid promptly and that trophies reach their winners. Organisers should also take time to note what worked well and what didn't, so that you can learn from this experience when you organise your next show.

For more detailed information on organising and running a show, consult the latest edition of the RHS's Horticultural Show Handbook.

Writing a Show Schedule

A schedule outlines everything there is to know about a show and provides a reference for exhibitors and judges. If you're writing a schedule for the first time, try contacting organisers of other local shows for advice. Here are a few pointers for what to include.

WHO'S WHO

This should include the names and contact details for the show secretary and any other relevant organisers. Judges' names should also be included, if available. Many societies also include the names of the society officers in their show schedules, even if these people are not directly involved in the show itself.

SHOW TIMETABLE

Include a timetable detailing:
- the deadline for the receipt of entries
- staging and judging times
- when the show opens to the public
- the deadline for any protests
- when prizes will be awarded
- when the show closes to the public
- when exhibits should be removed
- when prize money will be paid

A LIST OF ALL AVAILABLE CLASSES

It can be helpful to organise classes into sections. Small shows may simply divide classes along the lines of "fruit", "vegetables", "floral", and "homecraft" or "domestic". Larger shows may break these sections down further: for instance, "vegetables" could be broken down into "root vegetables" and "vegetables, other than root". Keep children's classes separate in a section of their own.

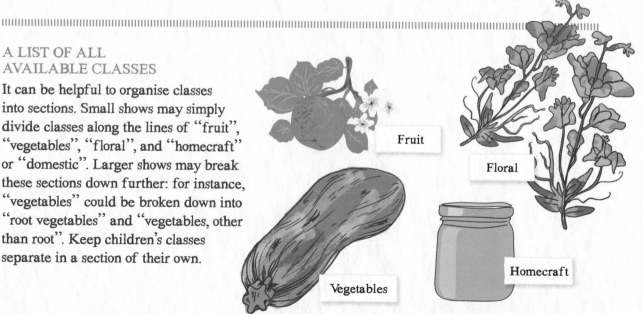

Fruit

Floral

Vegetables

Homecraft

Timings should be carefully considered, so that everything is ready when the public arrives

Proofread the draft schedule for any mistakes

Clearly state the range of produce permitted in collection classes

Consider having standard entry cards printed for exhibitors to fill in

ENTRY INFORMATION

Give full instructions about how to enter the show. Indeed, it can be helpful to include an entry form with the schedule itself, either printed on the final page or as a loose insert within the booklet. State the deadline for receipt of entry forms clearly on both the schedule and the form. The deadline should be set as close to show day as possible, to maximise the number of entries received, but be sure to allow enough time for all administrative work, such as arranging for enough table space and ordering staging materials, to be completed before show day. Some shows also charge exhibitors to enter the show, with fees typically modest; usually, this money goes towards the prize fund. If exhibitors will be charged, this should be clearly stated in the schedule. Children's classes are usually free of charge.

AWARDS AND PRIZE MONEY

Details of all prizes, including any special awards, should be clearly stated.

SHOW RULES

All rules that will be followed by the show should be clearly stated in full. This is for the benefit of both judges and exhibitors, as it allows the judging process to be as efficient and transparent as possible. See pp.214–15 for a standard set of rules, which can be amended for the needs of your particular show. If certain classes are to be judged according to the rules of a specialist national-level society, this should also be mentioned at this stage. Avoid including strict warnings for particular rules (e.g. "this rule will be strictly enforced"), as this implies that other rules will not be enforced.

Show Rules

The following rules have been adapted from those that apply to all Royal Horticultural Society competitions. They are suitable for many local shows and can be adapted for any particular needs you have.

1 Acceptance of Entries
The show committee reserves the right to refuse any entry and, in the event of such refusal, is not to be required to give any reason or explanation.

2 Eligibility of Exhibitors
On all questions regarding the eligibility of an exhibitor (e.g. whether they can be classed as a "novice"), the decision of the show committee shall be final.

3 Ownership of the Exhibits
All exhibits must be the property of the exhibitor (unless the exhibit is entered in joint names), and be the product of their own work.
Horticultural exhibits: All entries must have been grown from seed by the exhibitor, or have been in their possession or cared for by them for at least two months prior to the date of the show. This ruling does not apply to floral arrangement classes: in these instances, exhibitors are allowed to use plant material that they have not personally grown.
Domestic exhibits: Except as outlined in the schedule for children's classes, all entries must be the unaided work of the exhibitor. All entries must have been baked, or otherwise made, by the exhibitor, but the recipe need not be their own invention.

4 Number of Entries Allowed
Horticultural exhibits: Only one person per household may exhibit in a single class with produce from the same garden or allotment (unless the exhibit is entered into the class under joint names).
Domestic exhibits: No exhibitor may make and submit more than one entry per class.

5 Constitution of an Exhibit
Each exhibitor is responsible for compliance with the rules governing each class entered. If an error is noticed at any time, it may be corrected by the stewards, but they are under no obligation to do so.

6 Labelling Exhibits
All exhibits should be correctly and clearly labelled in block capitals. The use of correct and clear labelling may be viewed by the judges as an advantage in close competition.
Horticultural exhibits: Entry labels should state the correct name of the cultivar (variety). Errors in naming may not disqualify entries, except where the judges consider that an exhibitor is showing one cultivar under two or more names. If the cultivar name is unknown the label should read "NAME UNKNOWN". An unnamed seedling should be labelled "UNKNOWN SEEDLING".

Domestic exhibits: Labels should clearly describe the exhibit and indicate the main ingredients; usually, the full name of the recipe used will be sufficient. Any potential allergens (e.g. nuts, wheat) must also be listed on the label. All preserve labels must include the month and year the preserve was made.

7 Constitution of Fruit and Vegetable Dishes
Each dish must consist of one cultivar (variety) only. The numbers of specimens constituting dishes in a collection of fruit or vegetables must be those specified in the single-dish classes.

8 Layout of Exhibits
The show committee's officers may direct the placing of all exhibits.

9 Only One Prize in a Class
No exhibitor may be awarded more than one prize in any one class, unless specifically permitted by the schedule.

10 Prizes May Be Withheld
Prizes need not be awarded to exhibits considered to be below standard.

11 Exhibits Not According to Schedule
Any exhibit that does not conform to the wording of the schedule should be disqualified and a judge must write on the entry card "Not according to schedule" (NAS) in addition to a note as to why it is marked NAS. A vase/dish from a disqualified collection is not eligible for a best dish award.

12 Decisions
The decisions of the judges as to the merit of the exhibits shall be final. Any other points in dispute will be decided by the show committee and/or its appointed referees, particularly cases of disputed wording. In these cases, specimens may be withdrawn for further inspection.

13 Protests
Any protest must be made in writing and delivered to the secretary by the time stated in the schedule.

14 Alteration of Exhibits
After judging has taken place, no exhibit or part of an exhibit may be altered or removed until the end of the show, except by special permission of the secretary.

15 Liability for Loss
All exhibits, personal property, etc., will be at the risk of the exhibitors and the show committee will not be liable for compensation for loss or damage.

16 Right to Inspect Gardens of Exhibitors
In order to be satisfied that the conditions governing competitive horticultural exhibits are fulfilled, the show committee reserves the right for an official representative to visit by appointment, before or after any show, gardens from which plants, flowers, fruit, or vegetables have been entered for competition. If it is decided to exercise that right for one or more exhibitor, it does not mean that the gardens of any other exhibitors need be visited.

Glossary

All-round effect A term used to describe floral exhibits with an even, balanced distribution of flowers and foliage around the circumference of the plant or arrangement.

Alpine Any plant that is suitable for a rock garden or alpine house.

Bikini vase A green tapered vase, in standard sizes, used to display cut stems.

Blanch The part of any vegetable that is covered against sunlight.

Blemish Mark or imperfection on an exhibit that may be caused by mechanical damage, physiological deficiency, pest, or disease.

Bloom The waxy covering on many fruits and vegetables, as well as the leaves and stems of many succulents and other plants. Also used to refer to an open flower.

Bract Usually a small, leaf-like structure occurring below the flowers and above the true leaves. These can sometimes be coloured, as in *Euphorbia*.

Bulb An underground modified stem, bearing a number of swollen, fleshy leaf bases or scale leaves, which enclose the next year's bud.

Bulbous For horticultural-show purposes, "bulbous plants" describes those having bulbs, corms, or tubers. "Bulbous" may also refer to a defect, such as the swelling of a plant, e.g. the base of a spring onion.

Calyx The outer set of perianth segments, especially when green.

Class A subdivision of a competitive schedule; one group of comparable exhibits.

Collection An assembly of kinds and/or cultivars of plants, flowers, fruits, or vegetables in one exhibit.

Corolla The inner set of perianth segments, if differing from the outer set, and especially if coloured and showy.

Corona A trumpet- or cup-like development of the perianth.

Crumb The interior texture of a cake; the soft inner portion of baked bread, as distinguished from the crust.

Crust The hard, crunchy outside of baked bread, typically golden brown in colour. May also be used to refer to a tart or pie case.

Cultivar The internationally accepted term for what, in English-speaking countries, is commonly known by gardeners as a "cultivated variety" or simply a "variety".

Dish A specified number or quantity of a fruit or vegetable, constituting one item that may be displayed on a table, on a stand, or on another suitable receptacle. Unless permitted by the show schedule, a dish must consist of one cultivar only.

Display An exhibit in which attractiveness of arrangement and general effect are to be considered of more importance than they would have been had the schedule specified a "group" or a "collection".

Enriched In domestic classes, a bread dough made with butter, eggs, and sugar, or a cake that has been regularly infused with alcohol over weeks or months; fruit cakes are most commonly enriched.

Entry Both a notification of an intention to exhibit ("entry form") and a unit submitted for exhibition in a show.

Florets Small individual flowers, especially those in heads, as in a dahlia, chrysanthemum, or other members of *Asteraceae*. It also refers to individual clusters of edible flower heads, as in sprouting broccoli and calabrese.

Flower head For horticultural-show purposes, an assemblage of florets grouped together in a single head on a single flower stem (e.g. a disbudded chrysanthemum).

Foliage plant A plant usually grown for its ornamental foliage. If it is in flower, it may be entered into a foliage-plant class, but the flowers will not be taken into account.

Forced Grown to flower or be ready for consumption before the normal time.

Glaze This term can be used to describe the application of an egg wash on to raw dough or pastry just before baking, in order to enhance the colour. It may also refer to a thin citrus water icing poured over drizzle cakes, or a coating of warmed jam applied to the surface of a cake or tart.

Handle The portion of cucumber closest to the stem. This part will be less in girth than the main body of the fruit and tapering to the stem.

Hybrid A plant derived from the intercrossing of two or more genetically distinct plants.

Inflorescence The flowering portion of the stem above the last stem leaves, including its flower branches, bracts, and flowers.

NAS "Not According to Schedule": a note used by judges to describe exhibits that do not match the criteria specified in the schedule (for instance, a cake baked in a tin of an incorrect size).

Natural A lack of any artificial treatment, such as dyeing or varnishing.

Novice A competitor who has not won any prize at a previous show.

Originality In a schedule, "originality" means uncommon or unusual, but at the same time desirable.

Panicle A branched inflorescence.

Pectin The naturally occuring substance present in a variety of fruits at varying levels, which acts as a setting agent in the making of preserves.

Pedicel The stalk of a single flower on an inflorescence (see also Peduncle).

Peduncle The stalk of an inflorescence or of part of an inflorescence. This term should also be used for a stalk of an inflorescence with a solitary flower (see also Pedicel).

Perianth A term used for the calyx and corolla or their equivalents, but seldom used except when the segments of the two whorls are both coloured, as in a tulip.

Professional In show terms, a person who gains their livelihood by growing horticultural plants, flowers, fruit, or vegetables for sale, or for an employer or anyone employed in the maintenance of a garden, pleasure ground, or park.

Puff pastry A delicate, flaky pastry with multiple layers that "puff up" when baked. Sausage rolls are most commonly made with puff pastry.

Ray-florets The outer florets of a flower head, such as that of a daisy. They are often larger than the inner florets.

Rhizome An underground, usually horizontal, swollen stem containing food reserves, e.g. in bearded irises.

Rich See Enriched.

Rose end The end of a tuber where the dormant buds, or "eyes", are concentrated.

Seedling A young plant that has recently germinated.

Schedule A booklet issued for the use of a show's exhibitors, listing the rules for showing, as well as a list of classes and other relevant details (see pp.212–15 for more information).

Set In domestic terms, the thickness and consistency of preserves that form a gel or paste, such as jams, jellies, curds, and marmalade. A preserve's set should ideally be neither too firm nor too runny, although a jelly will typically have a softer set than a jam or marmalade.

Shortcrust pastry A relatively robust pastry used in both sweet and savoury bakes, such as pies, tarts, and quiches.

Spike For horticultural-show purposes, this is an unbranched (or only very slightly branched) inflorescence with an elongated axis, bearing either stalked or stalkless flowers, as in a gladiolus.

Spray A branched, many-flowered inflorescence on a single main stem.

Standard In horticulture, this term typically describes a specimen with an upright stem of some length, supporting a head. Roses, fuchsias, and chrysanthemums are ornamental plants readily grown as standards.

Strig A bunch of currants or hybrid berries. It is best detached from the plant with scissors and should not include any of the woody section at the base.

T-handle This is a feature of grape bunches, when the fruits are cut from a vine with a piece of lateral shoot either side of the stalk.

Tail The often curved, tapering end of a bean opposite to the stalk.

Truss A cluster of flowers or fruits growing from one main stem, as seen with tomato and pelargonium exhibits.

Tuber A swollen underground stem with buds or "eyes" from which new plants or tubers are produced.

Index

A

alpine plants 135
alpine strawberries 18
apples 36–39, 40
  cooking 37
  dessert 36–37
  dual-purpose 37
  unusual and heritage varieties 38–39
apricots 29
artichokes see globe artichokes; Jerusalem
  artichokes
asparagus 80
aubergines 62
auriculas 138

B

baking see bread; cakes; pastry; scones;
  shortbread
baking classes
  preparation 167
  presentation 167
  transit 167
beetroot 96, 113
  cylindrical 96
  globe 96
  long 96
beetroot cake 174, 175
beginners, tips for 112, 113
begonias 150
Bennett, Jeffery 158–59
blackberries 20–21
  hybrid berries 21
blackcurrants 24
bonsai 132–33
boysenberries 21
bread 187–89
  machine-made bread 187

broad beans 64
broccoli see calabrese; Romanesco broccoli;
  sprouting broccoli
Brooks, Richard 188–89
Brussels sprouts 115
bullaces 31

C

cabbages 110–11
  red 110, 111
  Savoy 110, 111
cacti 130–31
cakes
  children's bakes 177
  cupcakes/biscuits, decorated 177
  decorating cakes in the tent 167
  drizzle cakes 173
  fruit cakes 176
  vegetable cakes 172, 174, 175
  Victoria sandwich 168–69
calabrese 117
calamondin oranges 35
Cape gooseberries 50
capsicums see peppers
carnations 142–43
  border 142
  perpetual 142–43
carrots 94–95
  Chantenay 94
  long-pointed 94, 95
  Nantes 94
  stump-rooted 94, 95
cauliflowers 118
  coloured 118
cavolo nero 114
celeriac 90
celery 89
  self-blanching 89
  trench 89
Chant, Joy 180–81
chard 119
cheating 95, 113
cherries 28
  sour 28
  sweet 28
cherry plums 30
children
  children's bakes 177
  novelty classes 137

chilli peppers 59
chrysanthemums 160–61
chutneys 190, 193
citrus fruits 35
climbing beans see French beans
competitors
  beginners, tips for 112, 113
  bread baker 188–89
  cake baker 174–75
  camaraderie 113, 159
  cheating 95, 113
  competitive spirit 137
  curd maker 204–05
  dahlia exhibitor 158–59
  floral artist 128–29
  house plant exhibitor 136–37
  jam maker 200–201
  pumpkin exhibitor 78–79
  shortbread baker 180–81
  urban gardener 60–61
courgettes 72, 74
cucamelons 70, 71
cucumbers 70–71
  gherkin-type 70
cupcakes/biscuits, decorated 177
curds 202, 204–05
currants 24–25

D

dahlias 40, 156–59
damsons 31
decisions, judges' 215
delphiniums 144
domestic classes
  tips and tricks 166–67
  see also baking; preserves
drizzle cakes 173
dwarf beans see French beans

E

elephant garlic 87
entry information 213
exotic fruit and vegetables 41

F

Fear, Sue 204–05
fennel see Florence fennel
figs 34
floral displays 126, 128–29

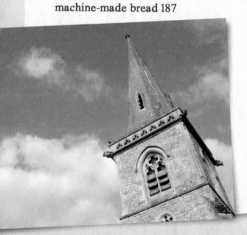

"all-round" effect 125
  arrangements 126, 128, 129
  mixed vases 126
Florence fennel 91, 120
flower classes
  cut stems 125
  floral displays 126
  potted plants 124
  tips and tricks 124-25
flowers, cut 125
  carnations 142-43
  chrysanthemums 160-61
  dahlias 40, 156-59
  delphiniums 144
  gladioli 151
  irises 140
  pinks 142-43
  preparation 125
  presentation 125
  roses 148-49
  sweet peas 145
  transit 125
  tulips 139
French beans 65
fruit
  exotic 41
  miscellaneous 48-51
  soft 18-27
  tree fruit 30-47
fruit cakes 176
fruit classes
  general classes 14, 19
  mixed collection and trug
    exhibits 15, 26
  polishing fruit 14
  single cultivar exhibits 14
  tips and tricks 14-15
fruit liqueurs 203
fruiting vegetables
  aubergines 62
  broad beans 64
  chilli peppers 59
  courgettes 72, 74
  cucamelons 70, 71
  cucumbers 70-71

French beans 65
  marrows 73, 74
  peas 67
  pumpkins 75-79, 200
  runner beans 66
  squashes 74, 76-77
  sweet peppers 58
  sweetcorn 63
  tomatoes 56-57, 113
fuchsias 152

G
gardens, inspection of 213
garlic 54, 87
  elephant garlic 87
giant fruit and vegetables
  cabbages 111
  carrots 94
  cucumbers 70
  gooseberries 27
  marrows 73
  onions 85
  Oriental radishes 104
  pumpkins 75, 77, 78-79
gladioli 151
globe artichokes 81
gooseberries 26-27
  red 27
grapes 48-49
green salad onions see spring onions

H
herbs 120-21
  cut herbs 120
  herb pots 120
house plants 127, 134, 136-37
  "all-round" effect 127
  flowering plants 127
  foliage plants 127
hybrid berries 21

I
insurance 211
irises 140

J
jams 198, 200-201
jellies 199
  sweet-savoury classes 199
Jerusalem artichokes 102
judges
  decisions 215
  finding 210
  impartiality 193, 210

marking 41
  RHS judges 40-41, 210
  skills and knowledge 193, 210
  stories 40-41, 192-93
  WI judges 192, 193, 204, 210

K
kale 114
  cavolo nero 114
  redbor kale 114
kiwi fruit 41
kohlrabi 97
kumquats 35

L
labelling exhibits 214
leafy vegetables
  Brussels sprouts 115
  cabbages 110-11
  calabrese 117
  cauliflowers 118
  chard 119
  kale 114
  lettuces 61, 106-07
  Romanesco broccoli 117
  spinach beet 119
  sprouting broccoli 116
leeks 82-83
  blanched 82, 83
  intermediate 82
  pot leeks 82, 83
lemon curd 202, 205
lemon tart 182-83
lemons 35
lettuces 61, 106-07
  butterheads 106, 107
  Cos 106
  crispheads 106, 107
  head-forming 106
  loose-leaf 106
Lloyd, Vivien 192-93
loganberries 21
Luckham, Sarahjane 174-75

M

mangetout peas 67
marmalade 196–97
marrows 73, 74
medlars 45
melons 41
mirabelles 30
mixed collection classes
  fruit 15
  preserves 166
  vegetables 55

N

novelty classes 137

O

Oliver, Matthew 78–79
onions 54, 84–85, 113
  pickling onions 84
  shallots 54, 60, 86
  spring onions 88
open classes 41
orchids 134, 136, 137
  tree-dwelling and rock-dwelling 134
organising a show 210–11
  categories and classes, choosing 210
  on the day 211
  entry information 211
  insurance 211
  officials, appointment of 210
  organisers' stories 112–13, 128–29, 204–05
  prizes 211, 212, 215
  publicity 211
  rules 211, 212, 214–15
  show secretaries 210
  stewards 210, 211
  venues 210
  writing a show schedule 212–15
  see also judges

P

Pacifico, Carol 136–37
pansies 141
  exhibition cultivars 141
  garden cultivars 141
parsnips 99
pastry
  lemon tart 182–83
  quiches 184–85
  sausage rolls 186
Pearce, Debbie 200–201
pears 42–43
peas 67
  mangetout 67
  sugar snap 67
pelargoniums 153
  ivy-leaved 153
  regal 153
  zonal 153
peppers
  chilli peppers 59
  sweet peppers 58
piccalilli 191
pink currants 25
pinks 142–43
Pizzoferro, Tony 112–13
Plescia, Sarah 60–61
plums 30
  cherry plums 30
  cooking plums 30
  dessert plums 30
  gages 30
  mirabelles 30
potatoes 103
  salad potatoes 103
potted plants 124
  "all-round" effect 125
  alpine plants 135
  auriculas 138
  begonias 150
  bonsai 132–33
  cacti 130–31
  fuchsias 152
  house plants 127, 134, 136–37
  irises 140
  orchids 134, 136, 137
  pansies 141
  pelargoniums 153
  preparation 124
  presentation 124
  primulas 138
  succulents 130–31, 136
  transit 124

preparation
  bakes 167
  cut flowers 125
  fruit 14, 15
  potted plants 124
  preserves 166
  vegetables 54, 55
  see also individual exhibits
presentation
  bakes 167
  cut flowers 125
  fruit 14, 15
  potted plants 124
  preserves 166, 193
  vegetables 54, 55
  see also individual exhibits
preserve classes
  mixed collection classes 166
  preparation 166
  presentation 166, 193
  transit 166
preserves
  chutneys 190
  curds 202, 204–05
  fruit liqueurs 203
  jams 198, 200–201
  jellies 199
  marmalade 196–97
  sweet pickles 190–91
  Women's Institute judges 192, 193, 210
primulas 138
prizes 211, 212, 215
publicity 211
pumpkins 75–79, 200

Q

quiche 184–85
  quiche lorraine 185
quince 44

R

radishes 104–05
  Oriental 104, 105
  salad 104, 105
  winter 104

raspberries 19
  hybrid berries 21
recipes
  apricot and passion fruit jam 201
  beetroot cake 175
  lemon curd 205
  lemon tart 183
  marmalade 197
  piccalilli 191
  quiche lorraine 185
  shortbread 181
  tear'n'share bread 189
  Victoria sandwich 169
redcurrants 25
red gooseberries 27
rhubarb 51
  categorising 51
Romanesco broccoli 117
roses 148-49
Royal Horticultural Society (RHS)
  Horticultural Show Handbook 40, 211
  RHS judges 40-41, 210
rules 211, 212, 214-15
runner beans 66

S

salad vegetables 91, 97, 98, 103, 110
sausage rolls 186
scones 178
shallots 54, 60, 86
shelling beans *see* French beans
shortbread 179-81
show schedule 212-15
show secretaries 210
soft fruit
  blackberries 20-21
  blackcurrants 24
  boysenberries 21
  currants 24-25
  gooseberries 26-27
  kiwi fruit 41
  loganberries 21
  melons 41
  pink currants 25
  raspberries 19
  redcurrants 25

strawberries 18
  tayberries 21
  white currants 25
  wineberries 21
sour gherkins *see* cucamelons
spinach beet 119
Spires, Colin 40-41
spring onions 88
sprouting broccoli 116
squashes 74, 76-77
staging *see* presentation
stems, bulbs, and roots
  asparagus 80
  beetroot 96, 113
  carrots 94-95
  celeriac 90
  celery 89
  Florence fennel 91, 120
  garlic 87
  globe artichokes 81
  Jerusalem artichokes 102
  kohlrabi 97
  leeks 82-83
  onions 84-85, 113
  parsnips 99
  potatoes 103
  radishes 104-05
  shallots 60, 86
  spring onions 88
  turnips 98
stewards 210, 211
stories
  cake baker 174-75
  curd maker 204-05
  dahlia exhibitor 158-59
  floral artist 128-29
  jam maker 200-201
  judges 40-41, 192-93
  preserves judge 192-93
  pumpkin exhibitor 78-79
  show organisers 112-13, 128-29, 204-05
  urban exhibitor 60-61
strawberries 18
succulents 130-31, 136
sugar snap peas 67
sweet peas 145
sweet peppers 58
sweet pickles 190-91, 193
sweetcorn 63

T

tayberries 21
timetable, show 212
tomatoes 56-57, 113

transit
  bakes 167
  cut flowers 125
  fruit 14, 15
  potted plants 124
  preserves 166
  vegetables 54, 55
tree fruit
  apples 36-39, 40
  apricots 29
  bullaces 31
  calamondin oranges 35
  cherries 28
  cherry plums 30
  citrus fruits 35
  damsons 31
  figs 34
  gages 30
  kumquats 35
  lemons 35
  medlars 45
  mirabelles 30
  plums 30
  quince 44
trug exhibitors
  fruit 15
  vegetables 55
tulips 139
turnips 98

U

urban shows 60-61

V

vegetable cakes 172, 174, 175
vegetable classes
  mixed collection and trug classes 55
  single cultivar exhibits 54
  staging 54
  tips and tricks 54-55
vegetables
  exotic vegetables 41
  fruiting vegetables 56-79
  herbs 120-21
  leafy vegetables 106-19
  stems, bulbs, and roots 80-105
Victoria sandwich 168-69

W

white currants 25
wineberries 21
Women's Institute judges 192, 193,
  204, 210

Matthew Biggs

Trained at The Royal Botanic Gardens at Kew, Matthew is a writer, lecturer, and television presenter. Affectionately known as "The People's Gardener", he is best known for being a regular panellist on BBC Radio 4's *Gardeners' Question Time*.

Thane Prince

Thane is the much-loved cookery writer and judge of BBC2's *The Big Allotment Challenge*. Her passionate expertise for preserves is unparalleled, and she has written multiple books on the subject, including *Thane Prince's Jams and Chutneys* (DK).

Acknowledgements

Thank you to Judy Palmer-Gowing for the judging criteria in the Homecraft Bakes & Preserves chapter, Marie Lorimer for indexing, and Emma Tennant for proofreading.

We are incredibly grateful for the warm welcomes and assistance received at the Lambeth Country Show, RHS Flower Show Tatton Park, Frome Agricultural and Cheese Show, Malvern Autumn Show, and RHS London Harvest Festival. Particular thanks go to Georgina Barter at the RHS; Rhiannon, Matilda, and Tony at the Lambeth Horticultural Society, and Tim at Lambeth Council; Maureen Hinton, Sue Fear, and the rest of the stellar team at the Frome Show; Jess Cook and Ellie Mainwaring at Silver Ball PR; Matthew Ford of HERS Agency.

Special thanks are extended to all our interviewees for sharing their stories, passions, tips, and recipes: Jeffery Bennett, Richard Brooks, Joy Chant, Sue Fear, Maureen Hinton, Vivien Lloyd, Sarahjane Luckham, Matthew Oliver, Debbie Pearce, Tony Pizzoferro, Sarah Plescia, and Colin Spires.

A final thank you to all the talented growers, flower arrangers, bakers, and preservers whose entries we photographed.

Picture Credits

The publisher would like to thank the following for their kind permission to reproduce their photographs:
(Key: b-below/bottom; l-left; r-right; t-top)
Alamy © Matthew Taylor/Alamy Live News: 8
Dorling Kindersley: Alan Buckingham: 31tr; 34br; 36bl; 46br; 50bl; 90br
Dorling Kindersley: Peter Anderson/National Dahlia Collection: 158bl
All other images © Dorling Kindersley.
For further information see: www.dkimages.com

DK | Penguin Random House

Senior Editor Alastair Laing
Editorial Assistant Amy Slack
Art Editors Mandy Earey, Anne Fisher, Vanessa Hamilton, Vicky Read, Jade Wheaton
Design Assistant Philippa Nash
Cover Design Nicola Powling
Pre-Production Producer Tony Phipps
Producer Stephanie McConnell
Managing Editor Stephanie Farrow
Managing Art Editor Christine Keilty
Art Director Maxine Pedliham
Publishing Director Mary-Clare Jerram

Photographer Mark Winwood
Additional Photography Nigel Wright, William Reavell
Heading Illustrations Steven Marsden

For Royal Horticultural Society
Editor Simon Maughan
Publisher Rae Spencer-Jones
Head of Editorial Chris Young

First published in Great Britain in 2017 in association with Royal Horticultural Society by Dorling Kindersley Limited
80 Strand, London, WC2R 0RL

The points tables and text for the judging notes and judging criteria in the chapters on Grow to Show Fruit, Vegetables, Flowers, and all text in the Nuts & Bolts chapter, are copyright © 2017 Royal Horticultural Society, based on material from the 8th edition of *The Horticultural Show Handbook*.

Foreword © 2017 Alan Titchmarsh.

A Penguin Random House Company
10 9 8 7 6 5 4 3 2 1
001–294007–Apr/2017

A CIP catalogue record for this book is available from the British Library.
ISBN 978-0-2412-5561-2

Printed and bound in China

A WORLD OF IDEAS:
SEE ALL THERE IS TO KNOW

www.dk.com